Meditations on Nature, Meditations on Silence

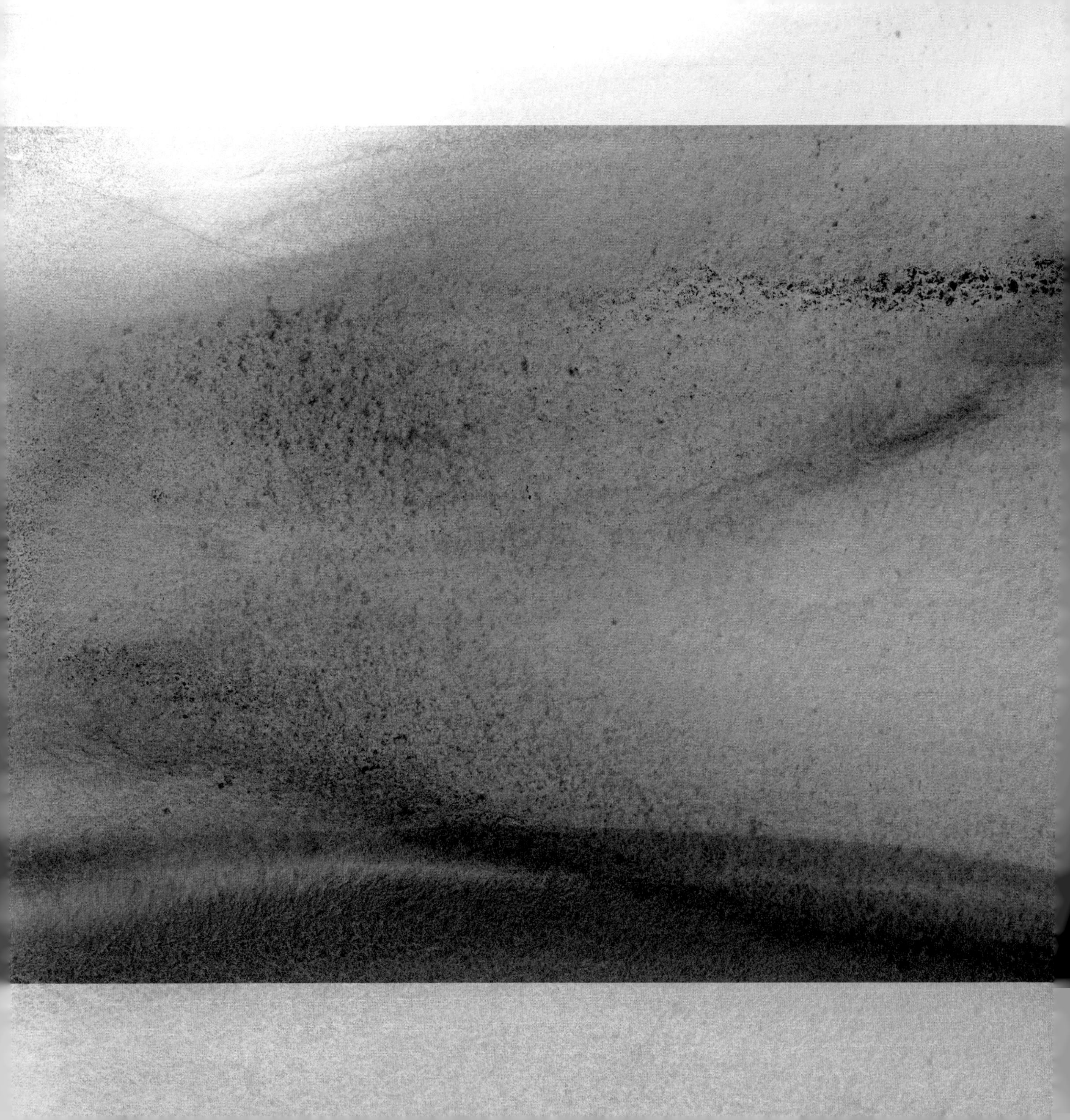

Meditations on Nature,
Meditations on Silence

Edited by Roderick MacIver and Ann O'Shaughnessy

HERON DANCE PRESS & ART STUDIO

450 Shunpike Road
Williston, VT 05495
888-304-3766
www.herondance.org

Printed in China by P. Chan & Edward, Inc.

Heron Dance donates art to dozens of grassroots wilderness protection groups each year. In addition, *Heron Dance* supports the Northeast Wilderness Trust with financial donations. For more information, please contact our office at 888-304-3766.

ISBN-10: 1-933937-11-4
ISBN-13: 978-1-933937-11-3

Design by Terry Fallon. Cover design by Luana Life.

For information about discounts for resellers, please contact Ingram Publisher Services at 800-961-7698 or visit www.herondance.biz.

We dedicate this book to Herb Pohl who drowned paddling on Lake Superior July 17, 2006.

Herb described himself as a romantic. He paddled all over Labrador and northern Canada. Most of his trips were solo. He loved wild places. He was an inspiration and a friend.

The Explorer

"There is no sense in going further — it's the edge of cultivation,"
So they said, and I believed it — broke my land and sowed my crop —
Built my barns and strung my fences in the little border station
Tucked away below the foothills where the trails run out and stop;

Till a voice as bad as Conscience, ran interminable charges
On one everlasting Whisper day and night repeated — so;
"Something hidden. Go and find it. Go and look behind the Ranges —
"Something lost behind the ranges. Lost and waiting for you. Go!"

—Rudyard Kipling, from "The Explorer"

I go out to wild places to experience the flow of natural life around me. What is important out there can't be named. It has to be felt. You can sense it when you step out of your car on a long drive through the prairies, you can sometimes hear it in the wind, but to really know it you've got to spend at least a few days out there and maybe a few weeks. It also helps if you can experience it alone.

Two weeks of paddling along the north shore of Lake Superior, as well as a number of trips that involved smaller, but still unpredictable, lakes in northern Canada, have taught me what it is like to live where the waves and wind are the dominant reality. Your little canoe is so vulnerable. You watch the clouds, the waves and try to intuit their meaning. In a photograph, those clouds and waves would be beautiful. Out on a big lake, they are more than that. Somehow the weather is like a volatile traveling companion whom you come to love, but never trust.

When you travel with a minimum barrier between yourself and nature, the land shapes your thinking. It invites you, and then forces you, to change your thinking. Eventually, awareness rather than conclusions and speech dominate your mind. You need a plan: in the worse case scenario, what will you do? Sometimes there is nothing you can do, and to avoid that circumstance, you study the horizon, the water, and wind.

And the physical exercise, hour after hour, day after day — in time your body comes to love that. You are in your body. It is your home, your friend. Gradually a feeling of peace and harmony with the land comes, and it is deeply satisfying. And the bugs — they are still there — all the time — but they only really bother you when you are tired. Maybe you forget about cooking dinner, grab some granola, some chocolate chips, eat in the tent and go to sleep. Maybe you get up in the middle of the night, break camp, and paddle, because the lake is flat and the going is good. Your sleeping, eating, and traveling have more to do with your Great Traveling Companion than with your watch. You forget about the agenda. Attunement to the land and water is perhaps the greatest satisfaction of a wilderness canoe trip. Gradually, you adopt the rhythm of the land and recognize it as your own. The few times I've experienced that have been among the most meaningful of my life.

I go to wild places to relax into the forces that swirl around me, mostly unseen. I go out there to experience the company of my subconscious mind, to avoid complexity and slow down. A few times in my life, perhaps a dozen, I've found a state of mind that might best be described as one of harmony and bliss. It seems to help if I've spent a few days exerting myself and experienced the cold, wet, or other discomforts. I've experienced it both alone and on trips with a partner, but it seems be more difficult when I am with others. Gradually my mind settles down. I know I'm where I hoped to be when I stop judging good and bad and descend into a kind of relaxed, pre-verbal awareness.

Once, towards the end of my first marriage, my then-wife packed up the kids and headed down to her family in the New York City area for the Columbus Day weekend. I threw my old canoe onto our other car, drove a couple of hours west, and headed up the Oswegatchie River in the Adirondacks' Five Ponds Wilderness. Three cold and rainy days later, I sat by the edge of the river brewing tea.

It was late afternoon and getting dark already, partly from the heavy cover of rain clouds, partly from the lateness of the season. I was sitting on my haunches, watching the river flow by. Somewhere, far off, a hermit thrush sang. I still remember its song: heavy and rich through dripping branches. The occasional trickle of water ran down the back of my neck and between my shoulder blades. I thought about putting up my tent, but I was enjoying the tea and was enjoying just sitting there listening and watching.

Two canoeists emerged out of the mist paddling hard upstream. It was the day before bow season started. They were the first *homo sapiens* I had seen since I started, and I had gotten used to the absence of talk and company and made no attempt to wave or say anything. They kept paddling, perhaps not noticing me, hunkered over as they were in their own discomfort and exertion. They disappeared around the bend, and it occurred to me right then that I had sunk into that quiet mind I crave.

That's roughly what happened. To some extent that experience, and the few others I've had like it, defy words. Early in a trip, the wanderer's mind is absorbed by what might go wrong, by fatigue, sore muscles, and sore feet. Gradually the mind and body adjust. The mind's incessant chatter dissipates.

Years ago, I interviewed Robert Perkins, creator of the film and author of the book, *Into the Great Solitude*, about his solo canoe trip down the Back River in the Canadian Arctic's Barren Lands. He described the process this way:

> I like the tundra because you can see a long way. You get shrunk to the right proportion in the expansiveness, when you are by yourself. Especially on longer trips. Somewhere in the middle, you can't reach backward, and you can't yet reach to the end. And there you are — just in the present moment. That is so exciting. I try to allow myself to flow into

the tundra around me. I try to hone my observation skills. Not only my eyes, but my ears; my whole body. I try to get sensitized to the creatures and the landscape. I have moved a long way into that world since I started.

During my first journeys, I was clunky and jittery. Even the wind hurt. Later, you come to the point when your body works well.

Somewhere during a trip, when you aren't as consumed with your own thoughts and your own fears, you begin to sense other stuff. Things about the animals and the land.

If you are with another person it is twice as hard. With five people, it is five times as hard. You end up tuning in to them, wanting to take care of them. Letting them take care of you. "Are you alright?" "Can I help?" Your mood is up. Mine is down. We are always looking in each other's eyes. If I am alone, I don't experience that. Instead, I have the sights and the sounds of nature. Of other creatures. They become companions.

Solitude is the deepest well I have ever come across. I imagine it would be different if solitude was forced on you, but to choose it is to find a source of sustenance that never runs out. It places a person in proper alignment, in proper order.... Every time I go on a solo canoe trip, I have to listen carefully to my thoughts and memories. It's the impact of stepping outside with a minimum of things and a great deal of landscape around you. A great deal of quiet. You begin to listen to what is around you and to what is going on inside of you.

It doesn't matter what your concerns have been over the past year — they just kind of boil off over the two months. Like maple syrup. You get down to some pretty fundamental, beautiful moments where you just catch yourself doing something. With no prior thought and no after-thought. You are totally absorbed making a fire, cooking dinner, or just paddling. Those moments are the reasons why I do those trips. I just love those moments.

I used to drive around the country interviewing people. I was partly on what I thought of as a search for truth, and partly trying to find people who would tell me what I wanted to hear. Mostly what I wanted to hear was that our culture was going down completely the wrong road, that a fulfilled, satisfying life is found in simple things, in adventure, in the creation of work of beauty.

One January, my travels brought me to northern Virginia — Amish country — and to the doorstep of folksinger Tom Wisner. Tom was in his early seventies and living in a former hippie commune. The hippies had gone out and gotten jobs, and now used the property for summer getaways. Tom was living in his car, traveling around the area performing his songs for a few bucks here and there. When one of the ex-hippies learned of his circumstances, she invited Tom to live on the property, and he set up in a big old farmhouse. I interviewed him there in a room lined with hundreds of books about the natural history of the Chesapeake River. Every so often, during the interview, he'd start strumming his guitar and singing. Mostly what he talked and sang about were the lives of the old time Chesapeake River fishermen.

I was working on yet another of my theories — that the solution to the fundamental problems of our culture was a love of natural beauty instead of an obsession with the material. Tom seemed the perfect person to try my theory out on. He responded (*Heron Dance* interview, Issue 20, Spring 2000):

> You use the word beauty a lot. I would encourage the word pace. Rhythm. For me, the word rhythm has emerged above all other words. The word beauty is kind of ephemeral for me. I think more about peace. I long for peace....

The people I have known on the water — the old-timers I knew — eased with the winds. Do you think they live high stress? No. They lived simple. Their bodies even moved with it. They knew a rhythm and a tempo — different from this hard-driving tempo we are into on the beltways. Wheels going to work. Drivin'.

I once watched an old guy adze out a helm. Adze it out. I watched his rhythm. I was blown away watching him. He worked that adze down that whole keel. Whack. Whack. For hours. When you talk to someone like that, or hang out with them, their whole rhythm of stories, of song, the way they move from song to story, is all in that beautiful simple balance. So I try to get hold of that in my song … and in the way I live.

I too have known people who lived close to the land and water. Their lives and speech have a deliberate and slow rhythm. They have tended to be people of few words. The beauty they've found in the woods is not a conversation topic that would come naturally to them. But they've also tended to be spiritual, and their spirituality manifested itself in how they live.

More than anything other than perhaps love, a life needs spirituality and it needs a rhythm, and those things have to work well together. Like humans, rhythms are obviously individual, but also like humans, rhythms bear a common ancestry — the wilderness from which we came. We lived out among the rivers and lakes, the animal migrations and the changing seasons, for well over a hundred thousand years. That is our real home, the ultimate repository of the human soul.

The rhythm of wild places has so little in common with the rhythm of our modern culture. I tell myself that I live in this culture, but am not of this culture. To some extent that's true. But I have bills, I run a business, I own a home and a car. I have three sons. Those moving parts and responsibilities demand a relationship with money, a relationship that imposes upon us a reality different from that found in the natural world.

It is a huge challenge to live a rhythm inspired by a close connection to the natural world in a culture that moves with an ever-increasing velocity. It requires some sacrifice and careful thought. It asks us to treat life as a precious gift.

—Roderick MacIver

Love all the Earth, every ray of God's light, every grain of sand or blade of grass, every living thing. If you love the Earth enough, you will know the Divine Mystery.

—*Fyodor Dostoevsky, from* The Brothers Karamazov

In orbiting the sun, the earth departs from a straight line by one-ninth of an inch every eighteen miles — a very straight line in human terms. If the orbit changed by one-tenth of an inch every eighteen miles, our orbit would be vastly larger, and we would all freeze to death. One-eighth of an inch? We would all be incinerated.

—Science Digest, *1981*

The miraculous is not extraordinary but the common mode of existence. It is our daily bread. Whoever really has considered the lilies of the field or the birds of the air and pondered the improbability of their existence in this warm world within the cold and empty stellar distances will hardly balk at the turning of water into wine — which was, after all, a very small miracle. We forget the greater and still continuing miracle by which water (with soil and sunlight) is turned into grapes.

—Wendell Berry

In the south, the red mountains fall away and yellow mountains rise up, full of silver and turquoise rock. There are plenty of rabbits here, a little rain in the middle of the summer, fine clouds tethered on the highest peaks. If you are out in the middle of the desert, this is the way you always end up facing.

In the south, twelve buckskin horses are living along the edge of the yellow mountains. The creeks here are weak; the horses have to go off somewhere for water, but they always come back. There is a little grass, but the horses do not seem to eat it. They seem to be waiting, or finished. Ten miles away you can hear the clack of their hooves against the rocks. In the afternoon, they are motionless, with their heads staring down at the ground, at the little stones.

At night they go into the canyons to sleep standing up.

From the middle of the desert, even on a dark night, you can look out at the mountains and perceive the differences in direction. From the middle of the desert, you can see everything well, even in the black dark of a new moon. You know where everything is coming from.

—Barry Lopez, from Desert Notes

The owl rides the meadow at his hunting hour. The fox clears out the pheasants and the partridges in the cornfield. Jupiter rests above Antares, and the fall moon hooks itself into the prairie sod. A dark wind flows down from Mandan as the Indians slowly move out of the summer campground to go back to the reservation. Aries, buck of the sky, leaps to the outer rim and mates with earth. Root and seed turn into flesh. We turn back to each other in the dark together, in the short days, in the dangerous cold, on the rim of a perpetual wilderness.

—*Meridel LeSueur, from* The Ancient People and the Newly Come

What is life? It is the flash of a firefly in the night. It is the breath of a buffalo in the winter time; it is the little shadow, which runs across the grass and loses itself in the sunset.

—*Last words of Crowfoot, Blackfoot hunter*

Instead of dashing through the day, completing one activity as quickly as possible so as to get at something else, it would be better to savor each part of the day as if this were your last day. Get as much out of the present moment as possible, from daybreak to bedtime. Even sleeping should be done well, even it one must wake up now and then to enjoy the night. When you eat, give each dish its full importance and extract its individual flavor.... Most everything we do deserves reverence, and a special setting. But one must live naturally, without pretense.

—*Harlan Hubbard, from* Payne Hollow Journal

Our life is an apprenticeship to the truth that around every circle another can be drawn; that there is no end in nature, but every end is a beginning; that there is always another dawn risen on the mid-noon, and under every deep a lower deep opens.

—*Ralphy Waldo Emerson, from* Essays: First and Second Series

I am glad that I shall never be young without wild country to be young in. Of what avail are forty freedoms without a blank spot on the map?

—*Aldo Leopold, from* Sand County Almanac

My heart leaps into my mouth at the sound of the wind in the woods.

—*Henry David Thoreau, undated journal entry*

We don't have plays and music and contact with sophisticated minds, and a round of social engagements. All we have are sun and wind and rain, and space in which to move and breathe. All we have are the forests, and the calm expanses of the lakes, and time to call our own. All we have are the hunting and fishing and the swimming, and each other.

We don't see pictures in famous galleries. But the other day, after a sleet storm that had coated the world with a sheath of ice, I saw a pine grosbeak in a little poplar tree. The setting sun slanted through a gap in the black wall of the forest, and held bird and tree in a celestial spotlight. Every twig turned to diamond encrusted-gold, and the red of the bird's breast glowed like a huge ruby as he fluffed his feathers in the wind. I could hardly believe it. I could only stand still and stare.

And then I repeated to myself again something that I once learned in the hope that it would safeguard me from ever becoming hardened to beauty and wonder. I found it long ago, when I had to study Emerson: "If the stars should appear one night in a thousand years, how men would believe and adore; and preserve for many generations the remembrance of the City of God which has been shown!"

—*Louise Dickinson Rich, from* We Took to the Woods

i thank you God for most this amazing
day; for the leaping greenly spirits of trees
and a blue true dream of sky; and for everything
which is natural which is infinite which is yes

— *e.e. cummings*

We live in a truly marvelous world. A really, really interesting, diverse, marvelous place to have visited for a short time. Some get more out of it than others. Some of us try to get to get more out of it than others.... I think you should make a conscientious effort to try. To be nosy. To look and to marvel. And not only to look but to see. Not only listen but hear on all different levels. It is indeed a marvelous world. Part of what makes it marvelous is our own kind. Part of what makes it incredibly marvelous to me are other than our own kind. I think it is important biologically to have them, but it's also important for my quality of life. I would not want to live in a world that had only people in it. I like snakes and frogs and creepy, crawly things and marvelous birds that can fly two hundred miles an hours and free my spirit....

—*Len Soucy, from Issue 1 of* Heron Dance

This curious world which we inhabit is more wonderful than it is convenient; more beautiful than it is useful; it is more to be admired and enjoyed than used.

— Henry David Thoreau, from an undated journal entry

Dear friend, all theory is gray,
And green the golden tree of life.

—Goethe, from Faust

If I Were God

for Anne Dellenbaugh, wilderness guide

If I were God, I would make a world exactly like this one. I love its inconsistencies, its contradictions. I love it that this river flows around stones and finds its own way. I love it that people are free, even to be selfish and to think they own beaches and mountaintops and have the right to keep the poor off them. I love it that things change, that the boundaries of nations and the fences of the rich get torn down sometimes. I love it that some people think we have many lifetimes while others think we have only this one. I especially love it that no one knows for certain, even if they think they do.

I love it that there are little clovers here in the grass beside me as I write, the same kind I have known all my life, and that this morning there was a bewildered-looking moose that I have not known at all standing in the mist at the edge of this river.

I love it that I am sixty years old, and my hair is gray and my hand against this white paper is showing age spots, and I am sitting on a wedge of land between a river and a stream on a Monday afternoon in July. I love it that I don't know exactly where I am, because it helps me to remember that I don't know exactly where Earth is in this galaxy, or where this galaxy is in this universe, or whether I have only this lifetime or many lifetimes. I love supposing this one is the only one, because it keeps me mindful of how precious everything is.

I love it that there are little clovers here in the grass beside me as I write, the same kind I have known all my life, and that this morning there was a bewildered-looking moose that I have not known at all standing in the mist at the edge of this river.

There is sweet dock mixed in with the clover at my feet. My mother told me that sweet dock makes good greens. My family knew things that poor people had to know, like what wild greens you can eat. Right now, I am learning things only rich people get to know, like how it is to take a canoe trip led by a brilliant wilderness guide. Enid and I have left behind all the students in the writing workshops we lead for low-income women: Corinna and Diane and Maryann and Evelyn and Robin and Teresa and Lynn and a dozen more who can't be here because they are poor in money. Kate has money, but she can't be here because she is poor in health. And yet, if I were God, I would make a world just like this one, where everyone enters bloody between the legs or through the cut belly of a woman; where nothing is for certain and there is so much to learn. I would make the world unfair as this world is unfair, because only in a world like this one is it possible that maybe the rich will take down their fences;

maybe the poor will get together and break the fences down; maybe those who know how to read will teach those who don't. Maybe the fed will feed the hungry. Maybe the lion will lie down by the lamb.

Maybe none of this will happen, but if I were making a world I'd want it to be complicated and unfair, a place where everything needs everything else, where if someone kills off all the wolves then the moose will get sick and die slow deaths because nothing eats them anymore. I don't understand it, but I want to be here on this wedge of land, on this canoe trip, trusting myself to a woman who knows what I don't about rivers and weather and human bodies. I love how she told me last year that she could read the river. She knows by the ripples on its surface what lies beneath and where to take her canoe. I love it that this year she is teaching me where to take a canoe, and how. I love it that she is teaching me to brush mosquitoes away gently, instead of sending them to the next life — which I'm not at all sure they will have. I love being in this body, in this world, in this time and place. It took me sixty years to get here, sixty years to like my body well enough to walk bare into a river in full sight of other women without shame, sixty years to trust my body enough to believe I can paddle for six hours and still lead a writing workshop when I get home.

I want to be fully here. Tonight I will sleep between two streams of water, under stars that move from I don't know where to I don't know where. Right now, two dragonflies on my thigh are giving me a demonstration of the proper mating technique when you are shaped like a small stick with wings.

—Pat Schneider, from The Sun Magazine, *February 1997*

Rain

I like the gentle rains of spring to fall on the dry land.
I like to rub your back and feel your hair.
I like to lie quietly — my skin to your skin and have you touch me and me touch you.
I like the gentle rains of spring to caress the land.
To these things I sing.

—James Eggert

Most birds have a poor sense of smell but very good color vision. Hummingbird flowers are most often red — the color to which birds' eyes are very sensitive — and less often orange and yellow. ...it is now known that birds' color vision is better than ours: They see all the colors we do and ultraviolet as well, which is invisible, or black, to us. Some flowers develop ultraviolet patterns when they are most fertile and secreting the most nectar, which may be a way of making themselves even more attractive to hummingbirds. ...Another advantage of red is that it is not perceived by insects — they will not

be attracted to red flowers and into competition with hummingbirds. Some hummingbird flowers open for only one day and secrete their nectar before dawn, when they are visited by hummingbirds, which are active at first light, before bees and other cold-blooded insects have had time to warm up and set out in search of food.

—*Robert Burton, from* The World of the Hummingbird

It is truly amazing that hummingbirds feeding in North American yards in spring and summer will spend the winter thousands of miles to the south. And then one day in the spring, the hummingbirds suddenly reappear, having come all the way back to exactly the same feeder or bed of flowers.... While it is not difficult to appreciate that large, powerful fliers such as geese, swans, cranes, and even ducks and shorebirds can make intercontinental flights, it is not quite so easy to comprehend that this twice-yearly routine is also followed by small birds such as sparrows and finches. As for hummingbirds — such tiny scraps of life and tied to a lifestyle that requires frequent energy-rich meals — it seems inconceivable.... Rufous hummingbirds nesting along the southern coast of Alaska regularly spend the winter in Mexico, some 2,000 miles away: a few stay in the very south of Texas and along the coast of the Gulf of Mexico. The calliope hummingbird travels much the same route and, at 0.1 ounce, is the smallest long-distance migrant in the bird world.

—*Robert Burton, from* The World of the Hummingbird

I say that it touches a man that his blood is sea water and his tears are salt, that the seed of his loins is scarcely different from the same cells in a seaweed, and that of stuff like his bones are coral made. I say that physical and biologic law lies down with him, and wakes when a child stirs in the womb, and that the sap in a tree, uprushing in the spring, and the smell of the loam, where the bacteria bestir themselves in darkness, and the path of the sun in the heaven, these are facts of first importance to his mental conclusions, and that a man who goes in no consciousness of them is a drifter and a dreamer, without a home or any contact with reality.

—*Donald Culross Peattie, from* Almanac for Moderns

Sometimes I think it is a falcon
Sometimes I think it is a song
Sometimes I think it is God

—*Rainer Maria Rilke*

Statistically, the probability of any one of us being here is so small that you'd think the mere fact of existing would keep us all in a contented dazzlement of surprise. We are alive against the stupendous odds of genetics, infinitely outnumbered by all the alternates who might, except for luck, be in our places.

Even more astounding is our statistical improbability in physical terms. The normal, predictable state of matter throughout the universe is randomness, a relaxed sort of equilibrium, with atoms and their particles scattered around in an amorphous muddle. We, in brilliant contrast, are completely organized structures, squirming with information at every covalent bond. We make our living by catching electrons at the moment of their excitement by solar photons, swiping the energy released at the instant of each jump and storing it up in intricate loops for ourselves. We violate probability, by our nature. To be able to do this systemically, and in such wild varieties of form, from viruses to whales, is extremely unlikely; to have sustained the effort successfully for the several billion years of our existence, without drifting back into randomness, was nearly a mathematical impossibility.

—*Lewis Thomas, from* The Lives of a Cell

Sometimes, when a bird cries out,
Or the wind sweeps through a tree,
Or a dog howls in a far-off farm,
I hold still and listen a long time.

My world turns and goes back to the place
Where, a thousand forgotten years ago,
The bird and the blowing wind
Were like me, and were my brothers.

My soul turns into a tree,
And an animal, and a cloud bank.
Then changed and odd it comes home
And asks me questions. What should I reply?

—*Hermann Hesse*

Then a great peace came over me...
and I seemed to hear the pines and the wind
and the rocky shores say to me, "You ... lover of the wild, are part of us...."

—*Sigurd F. Olson*

The moth moves across the porch, millimeter by millimeter, a brief stage of a longer journey of energy from the core of the sun to the table of the ants. Protons fuse at the center of the sun, releasing energy. The energy diffuses upward, taking several million years to reach the sun's surface, where it is released as heat and light. The light streaks across ninety-three million miles of space, reaching the Earth eight minutes later, where it falls upon the green leaves of plants. The plants store the energy as carbohydrates. A moth stops at a flower of a plant and sips the sugary nectar. It uses the nectar's stored energy for flight, reproduction, and building a body rich with organic compounds. The moth beats its brains out against my porch light and falls dead to the floor, where it is discovered by a scout of a colony of ants. The call is raised: "Food!" Now the rest of the colony arrives, at first in ones and twos, then en masse. A storm of purpose ignites their tiny brains. Humping their backs and fiddling their legs, they have a go at the moth. The moth drifts across the porch floor, taking the packaged energy longer to cross a few feet of painted boards than it took to travel from sun to Earth.

—*Chet Raymo, from* Natural Prayers

My favorite, a quote that is probably responsible for my conversion to a nature-based spirituality, is from Chief Dan George. I found nothing in scripture to help me at all. I had to leave all that behind. I found Thoreau and Muir and those people. Then I found some native American writing. This is the one that did it. Just nailed me. I have carried it in my wallet for years:

> The beauty of the trees, the softness of the air, the fragrance of the grass speaks to me. The summit of the mountain, the thunder of the sky, the rhythm of the sea speaks to me. The faintness of the stars, the freshness of the morning dewdrop on the flower speaks to me. The strength of fire, the taste of salmon, the trail of the sun, and the life that never goes away, they speak to me, and my heart soars.

That was the one. I folded that up and stuck it in my wallet.

You can walk the forest and do exactly the same thing. It may seem odd at first, but you get on speaking terms with everything. You do that with animals, and you do that with little bugs, and you do that with everything. It simply means you are aware, you are opening, you are noticing, you are standing with something. It is relationship. Community. That is what will save the world. It is communion that will save the world. Communion with beauty.

—Fritz Hull, founder of the Whidbey Institute, from Issue 35 of Heron Dance

I was sitting out back on my 33,000-acre terrace, shoeless and shirtless, scratching my toes in the sand and sipping on a tall iced drink, watching the flow of evening over the desert. Prime time: the sun very low in the west, the birds coming back to life, the shadows rolling for miles over rock and sand to the very base of the brilliant mountains. I had a small fire going near the table — not for heat or light but for the fragrance of the juniper and the ritual appeal of the clear flames. For symbolic reasons. For ceremony. When I heard a faint sound over my shoulder I looked and saw a file of deer watching from fifty yards away, three does and a velvet-horned buck, all dark against the sundown sky. They began to move. I whistled and they stopped again, staring at me. "Come on over," I said, "have a drink." They declined, moving off with casual, unhurried grace, quiet as phantoms, and disappeared beyond the rise. Smiling, thoroughly at peace, I turned back to my drink, the little fire, the subtle transformations of the immense landscape before me. On the program: rise of the full moon.

—*Edward Abbey, from* Desert Solitaire

I believe a leaf of grass is no less than the journey-work of the stars,
And the pismire is equally perfect, and a grain of sand, and the egg of the wren,
And the tree-toad is a chef-d'oeuvre for the highest,
And the running blackberry would adorn the parlors of heaven,
And the narrowest hinge in my hand puts to scorn all machinery,
And the cow crunching with depress'd head surpasses any statue,
And a mouse is miracle enough to stagger sextillions of infidels.

—*Walt Whitman, from "Song of Myself," as quoted in* Earth, My Likeness

The danger is to see and hear with the intellect instead of the senses, or rather with the intellect alone, instead of the intellect through the senses. Nothing is more perishable than our relations with the earth. It must be constantly renewed. Come in a house, think of something else, become absorbed in some work — it is gone. This communion is only possible when the mind is free. The body may be doing whatever else it wants to do.

—*Harlan Hubbard, from* Payne Hollow Journal

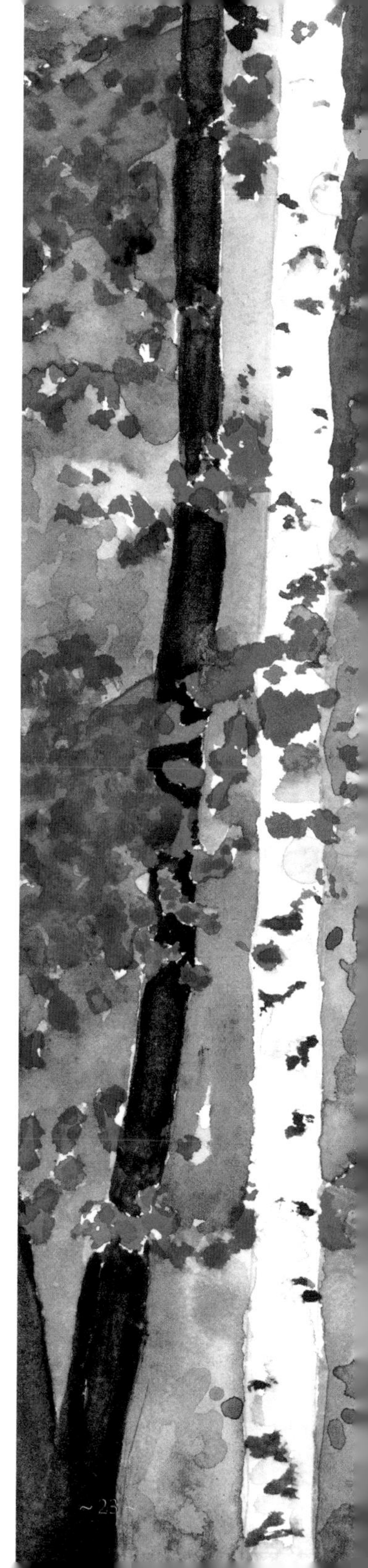

Dear Doris [Myers], August 30, Castle Crags

I have been feeling so happy and filled to overflowing with the beauty of life, that I felt I must write to you. It is all a golden dream, with mysterious, high, rushing winds leaning down to caress me, and warm and perfect colors flowing before my eyes. Time and the need of time have ceased entirely. A gentle, dreamy haze fills my soul, the rustling of the aspens lulls my senses, and the surpassing beauty and perfection of everything fills me with quiet joy and a deep pervading love for my world.

My solitude is unbroken. Above, the white, castellated cliffs glitter fairy-like against the turquoise sky. The wild silences have enfolded me unresisting.

Beauty and peace have been with me, wherever I have gone. At night, I have watched pale granite towers in the dim starlight, aspiring to the powdered sky, tremulous and dreamlike, fantastical in the melting darkness.

I have watched white-maned rapids, shaking their crests in wild abandon, surging, roaring, overwhelming the senses with their white fury, only to froth and foam down the current into lucent green pools, quiet and clear in the mellow sunlight.

On the trail, the musical tinkle of the burro bells mingles with the sound of wind and water, and is only heard subconsciously.

On the lake at night, the crescent moon gleams liquidly in the dark water, mists drift and rise like lifting enchantments, and tall, shadowed peaks stand guard in watchful silence.

These living dreams I wish to share with you, and I want you to know that I have not forgotten.

Love from Everett Ruess

—*Everett Ruess, from* Vagabond for Beauty

Once I swam down Washington's Duckabush River in a wet suit and mask. It was during the dog salmon run, and there was a flood of fish in the river. The current ran both ways that day. Halfway down the river I floated over a deep pool where an eddy had piled a pyramid of golden alder leaves. Further on, resting in the shallows and musing on what I'd seen, I noticed a shape move behind a submerged snag. It was a large male dog salmon, splotchy gray and yellow with faint copper tiger stripes, spawned out but alive in his eyes. I dove and glided toward him until we were a foot apart. I looked into his eye. He saw me but did not move. I was just another river shadow, an aspect of his dying, an aspect of his marriage, another guest at the feast. He was the eye of the resource, the subterranean sometime king, fish-eyed inscrutable god, alder-born-elder, tutor.

—*Tom Jay, from* Reaching Home: Pacific Salmon, Pacific People

Summer Night Storm

The ranting of the gods, this tumbling sky,
this wind-strong rain which pelts against my cheek,
the world re-lit by lightning, and the lie
of tall sea grass low bent against the sand.
I stand here, strangely still, with all the world
tumultuous at my feet, and yet my heart
is stronger than the roaring wind that swirls
about my body, taut against its force;
that blows my eyelids shut, that locks my lips,
lest all my spirit end its restlessness
in one wild song.

—*Jane Tyson Clement*

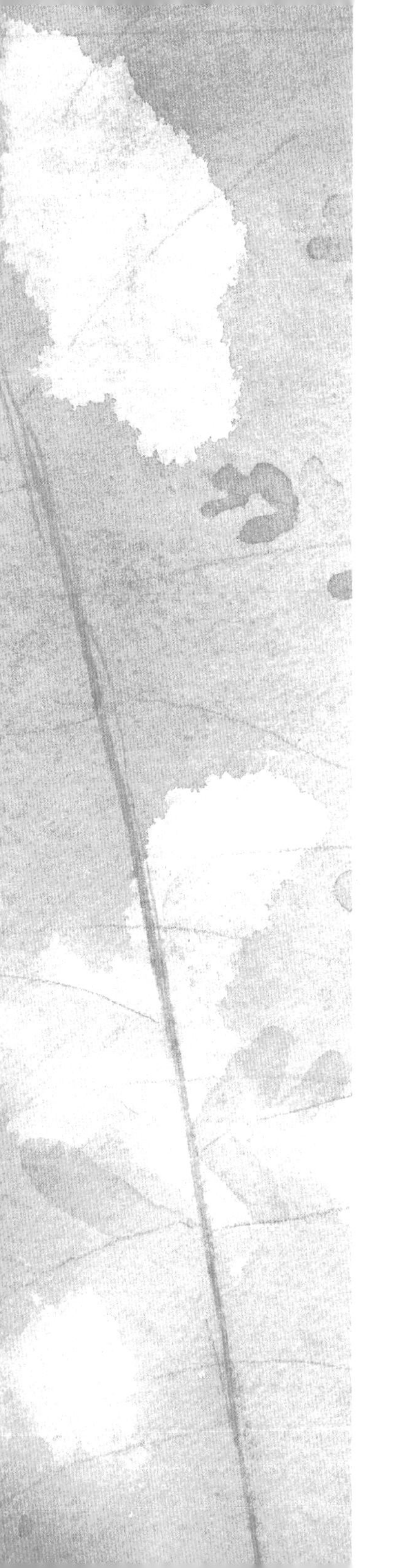

A reverence for life is a reverence for wildness. A reverence for life beyond your control. Something you don't dominate. That is the native habitat of new ideas. Of real humanity — to expose yourself to things beyond your control. And just ride out the consequences. That is what I seek and want to protect. Elements that are beyond our control.

—*Doug Peacock, from Issue 22 of* Heron Dance

Smell is the mute sense, the one without words. Lacking a vocabulary, we are left tongue-tied, groping for words in the sea of inarticulate pleasure and exaltation. We see only when there is light enough, taste only when we put things into our mouths, touch only when we make contact with someone or something, hear only sounds that are loud enough. But the smell is always and with every breath.... Etymologically speaking, a breath is not neutral or bland — it's cooked air; we live in a constant simmering. There is a furnace in our cells, and when we breathe we pass the world through our bodies, brew it lightly, and turn it loose again, gently altered for having known us.

—*Diane Ackerman, from* A Natural History of the Senses

Drinking coffee with honey in it and canned milk, smoking a pipe that had the sweetness pipes only have in cold quiet air, I felt good if a little scratchy-eyed, having gone to sleep the night before struck with the romance of stars and firelight, with the flaps open and only the blanket over me, to wake at two-thirty chilled through.

On top of the food box, alligator-skin corrugations of frost had formed, and with the first touch of the sun the willows began to whisper as frozen leaves loosed their hold and fell side-slipping down through the others that were still green. Titmice called and flickers and a redbird, and for a moment, on a twig four feet from my face, a chittering kinglet jumped around alternately hiding and flashing the scarlet of its crown.... I sat and listened and watched while the world woke up, and drank three cups of the syrupy coffee, better I thought than any I'd ever tasted, and smoked two pipes.

You run a risk of thinking yourself an ascetic while you enjoy, with that intensity, the austere facts of fire and coffee and tobacco and the sound and feel of country

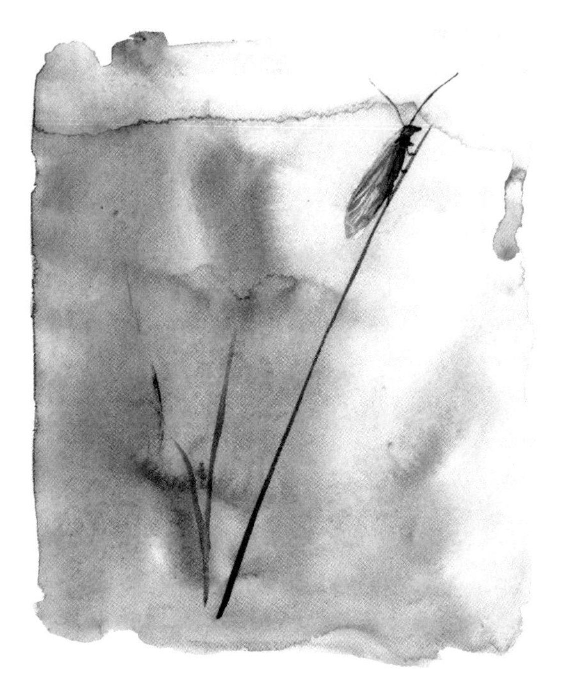

places. You aren't, though. In a way you're more of a sensualist than a fat man washing down sauerbraten and dumplings with heavy beer while a German band plays and a plump blonde kneads his thigh.... You've shucked off the gross delights, and those you have left are few, sharp, and strong. But they're sensory. Even Thoreau, if I remember right a passage or so on his cornbread, was guilty, though mainly he was a real ascetic.

—*John Graves, from* Goodbye to a River

As the conversation turned to waiting, Brother Anthony leaned forward in his chair. "Contemplative waiting is consenting to be where we really are," he explained. "People recoil from it because they don't want to be present to themselves. Such waiting causes a deep existential loneliness to surface, a feeling of being disconnected from oneself and God. At the depths there is fear, fear of the dark chaos within ourselves."

Brother Anthony was exactly right. Ultimately, we are fleeing our own dark chaos. We are fleeing ourselves.

—*Sue Monk Kidd, from* When the Heart Waits

The truth is that we need invertebrates but they don't need us. If human beings were to disappear tomorrow, the world would go on with little change. Gaia, the totality of life on Earth, would set about healing itself and return to the rich environmental states of 100,000 years ago. But if invertebrates were to disappear, it is unlikely that the human species could last more than a few months. Most of the fishes, amphibians, birds, and mammals would crash to extinction about the same time. Next would go the bulk of the flowering plants and with them the physical structure of the majority of the forests and other terrestrial habitats of the world. The soil would rot. As dead vegetation piled up and dried out, narrowing and closing the channels of the nutrient cycles, other complex forms of vegetation would die off, and with them the last remnants of the vertebrates. The remaining fungi, after enjoying a population explosion of stupendous proportions, would also perish. Within a few decades the world would return to the state of a billion years ago, composed primarily of bacteria, algae, and a few other very simple multicellular plants.

—*Edward O. Wilson, from* In Search of Nature

Secret Pond by Ann O'Shaughnessy

I am crouching still near a tree on a loamy ridge, my two hands spread around the trunk. I am feeling grateful for this tree that I remember because of its mossy smell and thick crevassed bark. It tells me that the beaver pond is near where one white pine shoots 100 feet up out of the tannic water, which means I am close to camp and food and sleep.

I get to the pond's edge, across from the point where my tent sits. There are no trails and the boreal forest is thick with scrub pine and dead-fall. Early afternoon sun brings out a wave of deer flies; one bites into my back and prickers dig into my shins. I can see my tent across the pond, 100 yards as the crow flies, but probably a mile of rough going around the edge. I decide to take off my clothes, leave them on this rock by the shore, swim across and come back for my things later in my canoe. Even though the whine of the deer flies' wings beating around my head intensifies, I just stare at the water. A large fallen pine tree leads from the shore and disappears into the darkness. A fear takes hold of me, as it does every time I contemplate swimming in this dark water.

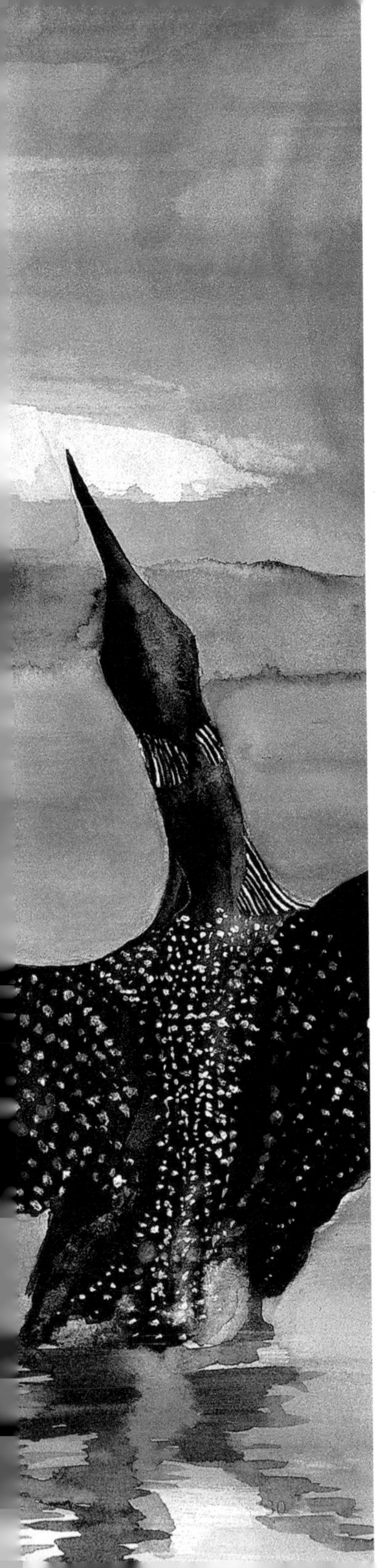

I shake my head to loosen its grip, feel a deer fly land on the small of my back and I dive. I swim as hard as I can, my heart banging away, my eyes closed. After twenty strokes, I remember the old pine and open my eyes afraid I will hit it. But I am far past it, and now I do the trick that always relieves my fear: as I lift my arm to pull through the water, I watch the bubbles as they leave my nose. I concentrate on my breath. Soon I am laughing at myself. Fear is at once humbling and freeing.

Two hundred strokes out I stop and tread water to look around. I have outraced the deer flies and now I see a pair of loons swimming nervously back and forth twenty yards away. I must be close to their nest. This is my second day, I hope by the end of the week they will have relaxed with me in their home.

I force myself to look down at the water that hides my naked body. I am trying not to think of my white toes dangling five feet down, looking like fish bait or worse — snapping turtle bait. As a kid, I thought something I could not see would grab my toe and drag me under. But still I was always drawn by the orange-brown color and rich-with-life smell. It is the smell of fishing with dad and my brother and the smell of excursions with my brother and sister in our old row boat deep into the swamp a mile from our house. I love to recall the easy silence between us as we drifted farther and farther back into the swamp away from the sounds of the road. Back there other sounds took hold of us: the quiet creaking of the oar lock; the turtle's claws scratching as it scurried off the fallen, half-submerged tree; the song of the red-winged blackbird as it announced its territory to us. Young as we were, our souls opened easily to those sounds and the almost-sweet smells of decay, connecting us to what is nameless and peaceful and old inside. Fantasies came to me there — of being a brave explorer deep in a northern wilderness. I would feed that feeling, and my brother and I would exchange the looks of bold adventurers. Back in the swamp we never bickered; we just listened, were silent with wide eyes, and felt as if we were part of something important. Today, the velvet water cleans away the sweat and the blood, soothes the deep scratches. I feel like an Amazon woman.

Near the edges, leaches like to wait for warm bodies, so when I reach the other side I scurry up the slippery rock and stand to look across to where I had started. The loons are quiet now, gliding back to their nest. I close my eyes and feel the sun soothing my worn muscles. I am aware of my body. I look down at the scratches and bruises, the big and small scars. I look at my muscular "boy" body and the breasts that nursed two babies. I think of how I tell my six-year-old daughter that scars are great stories that never fade — reminders of our play and adventures. I feel the ache in my shoulders from yesterday's portage and recall my dad firmly placing his hand on one shoulder when I was eight and saying, "Good solid girl." I love that my dad taught me how to hammer asphalt roofing, how to make a fireplace poker in his forge, how to tear down a wall and build a new one. I love that he celebrated the Amazon in me.

Today I am smiling wide and grateful for this body that hoisted the laden canoe over five beaver damns to get me here to this beautiful place. The same body that rose early to see the sun rise.

I lie down on the warm rock at the edge of the pond, and I close my eyes. My breath feels easy and light, my belly is soft and where a hard gnarled knot used to be under my sternum, a warmth spreads beyond my skin, around the blue sky and sun and back in again.

I need to recover a rhythm in my heart that moves my body first and my mind second, that allows my soul to catch up with me. I need to take a sacred pause, as if I were a sun-warmed rock in the center of a rushing river.

—*Dawna Markova, from* I Will Not Die an Unlived Life

For All

Ah to be alive
on a mid-September morn
fording a stream
barefoot, pants rolled up
holding boots, pack on,
sunshine, ice in the shadows,
northern rockies.

Rustle and shimmer of icy creek waters
stones turn underfoot, small and hard on toes
cold nose dripping
singing inside
creek music, heart music
smell of sun on gravel.
I pledge allegiance.

I pledge allegiance to the soil
of Turtle Island
one ecosystem
in diversity
under the sun—
With joyful interpenetration for all.

—*Gary Snyder, from* Axe Handles

Water has seen it all, its molecules have been through it all, and yet water silently continues to hold the secrets of its immense past. Perhaps the next raindrop once moistened the eye of a browsing dinosaur. Wouldn't it be amazing to be able to know the entire history of a single drop of water, a bead of sweat, or a tear?

—Patricia Barber, from Water Music, *photographed and orchestrated by Marjorie Ryerson*

Long-range migrations fascinate us. We are in awe of the distances salmon travel and their seemingly miraculous return to the exact point on earth where they emerged from their egg sacs to become fish. But salmon are of nature, and though our understanding of them remains incomplete, we know they do not require navigational miracles to reach home. Their in-stream and near shore courses are probably set by a combination of rheotaxis — detection of the direction of flow — and their ability to sense temperature changes and combinations of smells from distinct watersheds. Their sense of smell is thousands of times more acute than that of dogs. A salmon can probably detect one part per trillion by smell, or, in martini equivalents, roughly one drop of vermouth in 500,000 barrels of gin.

At sea, during voyages that can be 10,000 miles long, the salmon do not aimlessly wander, nor do they leave the navigation to their prey by simply following their next meals. They clearly orient themselves in some way, and swim homeward with precision equaling electronically equipped ocean sailors. Celestial, solar, and sonar explanations for the salmon's ability to navigate have been explored and, generally, dismissed. Most recently, fish thinkers at the University of Washington — long a mecca for salmon science — proposed an electro-magnetic solution to the puzzle of oceanic salmon migration. The earth's magnetic field produces an infinitely divided, arching grid of extremely low-voltage currents, so a salmon or any other creatures capable of detecting that voltage could track an arc on the grid that would lead it back to its starting point on the coast.

There, a salmon tunes its senses for local navigation by smell or rheotaxis to guide it to the patch of streambed or lake bottom from which it emerged many months earlier. Then, shocked by the freshwater of its birth, the beautiful swimmer transforms itself from a graceful silver voyager into a humpbacked reproductive aggressor, its bright ocean sheen replaced by the mottled combat of bruises to disguise itself in the shallows and present a fierce visage to competitors. The salmon passes a tower, a tallywhacker clicks, a biologist swats a mosquito, the journey ends, the journey begins.

—*Tom Jay and Brad Matsen, from* Reaching Home: Pacific Salmon, Pacific People

I watched the sun go down; the deserted island opposite me glowed rosily, happily like a cheek after a kiss. I heard the small songbirds returning drowsily to go to sleep, tired after a full day's hunting and singing. Soon the stars would rise to take their places one by one, and the wheel of night would begin to turn. Midnight would come, dawn would come, the sun would assuredly appear, and the wheel of day would commence its round.

A divine rhythm. Seeds in the ground, birds, stars — all obey. Only man lifts his hand in rebellion and wants to transgress the law and convert obedience into freedom. This is why he alone of all God's creatures is able to sin. To sin — what does that mean? It means to destroy harmony.

—*Nikos Kazantzakis, from* Report to Greco

There are billions of times more stars in our universe than there are grains of sand on our Earth.... There are on average about 100 billion stars more or less like our sun in each galaxy. In the visible universe there are perhaps 100 billion galaxies. Each one is at least several thousand light years across.... The closest galaxy similar to our own is more than 1 million light years away from us.... Scattered about the universe are huge voids, regions 10 to 30 million light years across, containing essentially no visible galaxies. There also exist enormous clusters of hundreds or thousands of galaxies, containing up to a million billion stars.... Galaxies continue to be observed out to the furthest reaches of the visible universe, more than one billion light years away.

—*Lawrence M. Krauss, from* The Fifth Essence: The Search for Dark Matter in the Universe

Nancy Newhall wrote: "Wilderness holds answers to questions man has not yet begun to ask." Some of these questions are scientific or ecological, some of them spiritual.

—*Bill Devall, from* Simple in Means, Rich in Ends: Practicing Deep Ecology

To those who followed Columbus and Cortez, the New World truly seemed incredible because of the natural endowments. The land often announced itself with a heavy scent miles out into the ocean. Giovanni di Verrazano in 1524 smelled the cedars of the East Coast a hundred leagues out. The men of Henry Hudson's Half Moon were temporarily disarmed by the fragrance of the New Jersey shore, while ships running further up the coast occasionally swam through large beds of floating flowers. Wherever they came inland they found a rich riot of color and sound, of game and luxuriant vegetation. Had they been other than they were, they might have written a new mythology here. As it was, they took inventory.

—*Frederick Turner*

Even in the worm that crawls in the earth there glows a divine spark.

—*Isaac Singer*

I think the value of the game of identification depends on how you play it. If it becomes an end in itself I count it of little use. It is possible to compile extensive lists of creatures seen and identified without ever once having caught a breath-taking glimpse of the wonder of life. If a child asked me a question that suggested even a faint awareness of the mystery behind the arrival of a migrant sandpiper on the beach of an August morning, I would be far more pleased than by the mere fact that he knew it was a sandpiper and not a plover.... Those who dwell, as scientists or laymen, among the beauties and mysteries of the earth are never alone or weary of life. Whatever the vexations or concerns of their personal lives, their thoughts can find paths that lead to inner contentment and to renewed excitement in living. Those who contemplate the beauty of the earth find reserves of strength that will endure as long as life lasts. There is symbolic as well as actual beauty in the migration of the birds, the ebb and flow of the tides, the folded bud ready for spring. There is something infinitely healing in the repeated refrains of nature — the assurance that dawn comes after night, and spring after the winter.

—*Rachel Carson, from* The Sense of Wonder

I finished my walk on the forest's edge, where the great music of crashing waves flooded into the tide pools, where wind ruffled devil's club leaves and hermit thrushes sang. I reminded myself that the wisest, most inspired people I knew had all taken this second path, heading for what I call the Far Outside. It is the path found when one falls into "the naturalist's trance," the hunter's pursuit of wild game, the curandera's search for hidden roots, the fisherman's casting of the net into the current, the water witcher's trust of the forked willow branch, the rock climber's fixation on the slightest details of a cliff face. Why is it that when we are hanging from the cliff — beyond the reach of civilization's safety net, rather than in it — we are most likely to gain the deepest sense of what it is to be alive? Arctic writer-ethnographer Hugh Brody has brooded over this question while working in the most remote human communities and wildest places he can find. There, he admits, "at the periphery is where I can come to understand the central issues of living."

—*Gary Paul Nabhan, from* Cultures of Habitat

Life must breed. Nature has no use for organisms, variations, or groups that cannot reproduce abundantly. She has a passion for quantity as prerequisite to the selection of quality; she likes large litters, and relishes the struggle that picks the surviving few; doubtless she looks on approvingly the upstream race of a thousand sperms to fertilize one ovum.

—*Will and Ariel Durant, from* The Lessons of History

Country Cottage

A peasant's shack beside the
Clear river, the rustic gate
Opens on a deserted road.
Weeds grow over the public well.
I loaf in my old clothes. Willow
Branches sway. Flowering trees
Perfume the air. The sun sets
Behind a flock of cormorants,
Drying their black wings along the pier.

—*Tu Fu, translated by Kenneth Rexroth,*
One Hundred Poems from the Chinese

I did not read books the first summer; I hoed beans. Nay, I often did better than this. There were times when I could not afford to sacrifice the bloom of the present moment to any work, whether of the brain or the hands. I love a broad margin to my life. Sometimes, in a summer morning, having taken my accustomed bath, I sat in my sunny doorway from sunrise till noon, rapt in a reverie, amidst the pines and hickories and sumacs, in undisturbed solitude and stillness, while the birds sang around or flitted noiseless through the house, until by the sun falling in at my west window, or the noise of some traveler's wagon on the distant highway, I was reminded of the lapse

of time. I grew in those seasons like corn in the night, and they were far better than any work of the hands would have been. They were not time subtracted from my life, but so much over and above my usual allowance. I realized what the Orientals mean by contemplation and the forsaking of works. For the most part, I minded not how the hours went. They day advanced as if to light some work of mine; it was morning, and lo, now it is evening, and nothing memorable is accomplished. Instead of singing like the birds, I silently smiled at my incessant good fortune. As the sparrow had its trill, sitting on the hickory before my door, so had I my chuckle or suppressed warble which he might hear out of my nest. My days were not days of the week, bearing the stamp of any heathen deity, nor were they minced into hours and fretted by the ticking of a clock; for I lived like the Puri Indians, of whom it is said that "for yesterday, to-day, and to-morrow they have only one word, and they express the variety of meaning by pointing backward for yesterday, forward for tomorrow, and overhead for the passing day." This was sheer idleness to my fellow-townsmen, no doubt; but if the birds and flowers had tried me by their standard, I should not have been found wanting. A man must find his occasions in himself, it is true. The natural day is very calm, and will hardly improve his indolence.

—*Henry David Thoreau, from* Walden

Where nature is concerned, familiarity breeds love and knowledge, not contempt.

—*Stewart L. Udall*

People often asked him which of all the creatures encountered in his many years as a hunter and dweller, in far-away places of Africa, he found most impressive. Always he answered that it would have to be a bird of some kind. This never failed to surprise them, because people are apt to be dazzled by physical power, size, frightfulness, and they expected him to say an elephant, lion, buffalo or some other imposing animal. But he stuck to his answer; there was nothing more wonderful in Africa than its birds. I asked why precisely. He paused and drew a circle with his finger in the red sand in front of him before saying that it was for many reasons, but in the first place

because birds flew. He said it in such a way that I felt I had never before experienced fully the wonder of birds flying.

I waited silently for him to find the next link in his chain of thought. In the second place, he remarked, because birds sang. He himself loved all natural sounds in the bush and the desert, but he had to admit none equaled the sounds of birds. It was as if the sky made music in their throats and one could hear the sun rise and set, the night fall and the first stars come out in their voices. Other animals were condemned to make only such noises as they must, but birds seemed free to utter the sounds they wanted to, to shape them at will and invent new ones to express all the emotions of living matter released on wings from its own dead weight. He knew of nothing so beautiful as the sight of a bird utterly abandoned to its song, every bit of its being surrendered to the music, the tip of the tiniest feather trembling like a tuning fork with sound. Sometimes, too, birds danced to their own music. And they not only sang.

They also conversed. There appeared to be little they could not convey to one another by sound. He himself had always listened with the greatest care to bird sound and never ceased to marvel at the variety of intelligence it conveyed to him.

Stranger still was their capacity of being aware of things before they happened. This was positively amazing. When the great earth tremor shook the northern

Kalahari some years before, Ben was traveling with a herd of cattle along the fringes of the Okovango swamp. One day he was watching some old-fashioned storks, sacred ibis, and giant herons along the edges of a stream. Suddenly the birds stopped feeding, looked uneasily about them, and then all at once took to their wings as if obedient to a single command. They rose quickly in the air and began wheeling over the river, making the strangest sounds. The sound had not fallen long on the still air before the ground under his feet started to shake, the cattle to bellow and run, and as far as his eyes could see the banks of the stream began to break away from the bush, as if sliced from it by a knife, and to collapse into the water. He had no doubt the birds knew what was coming, and he made a careful note of their behavior and the sound they uttered.

Even more wonderful, however, was their beauty. Color, for instance, lovely as it was in most animals, served the latter only for camouflage. But with birds it was much more. Of all the creatures, none dressed so well as the birds of Africa. They had summer and winter dresses, special silks for making love, coats and skirts for travel, and more practical clothes that did not show the dirt and wear and tear of domestic use. Even the soberest ones among them, which went about the country austere as elders of the Dutch Reformed Church collecting from parsimonious congregations on Sunday mornings — the old-fashioned storks in black and white, or the secretary birds with their stiff starched fronts and frock coats — their dress was always of an impeccable taste.

... Finally, there was their quality of courage. When one considered what tender, small, delicate, and defenseless things most birds were, they were perhaps the bravest creatures in the world. He had seen far more moving instances of the courage of the birds of Africa than he could possibly relate, but he would mention only one of the most common — birds defending their nests against snakes. On those occasions, they had a rallying cry, which was a mixture of faith and courage just keeping ahead of despair and fear. It would draw birds from all around to the point of danger, and the recklessness with which one little feathered body after another would hurl itself at the head of a snake, beating with its wings and shrieking its Valkyrian cry, had to be seen to be believed. Ben once saw a black mamba driven dazed out of a tree by only a score or so of resolute little birds. The mamba, which he killed, measured close on ten feet, and this snake is itself a creature of fiery outage and determination. No, all in all, he had no doubt that birds were the most wonderful of all living things.

—*Laurens van der Post, from* The Heart of the Hunter

Animal play is a reasonably common phenomenon, at least among certain mammals, especially in the young of those species. Play activities — by definition — are any that serve no immediate biological function, and which therefore do not directly improve the animal's prospects for survival and reproduction. The corvids, according to expert testimony, are irrepressibly playful. In fact, they show the most complex play known to birds. Ravens play toss with themselves in the air, dropping and catching again a small twig. They lie on their backs and juggle objects (in one recorded case, a rubber ball) between beak and feet. They jostle each other sociably in a version of "king of the mountain" with no real territorial stakes. Crows are equally frivolous. They play a brand of rugby, wherein one crow picks up a white pebble or a bit of shell and flies from tree to tree, taking a friendly bashing from its buddies until it drops the token. And they have a comedy-acrobatic routine: allowing themselves to tip backward dizzily from a wire perch, holding a loose grip so as to hang upside down, spreading out both wings, then daringly letting go with one foot; finally, switching feet to let go with the other. Such shameless hot-dogging is usually performed for a small audience of other crows.

—*David Quammen, from "Has Success Spoiled the Crow?"* from Natural Acts

I have always longed to be a part of the outward life, to be out there at the edge of things, to let the human taint wash away in emptiness and silence as the fox sloughs his smell into the cold unworldliness of water; to return to the town as a stranger. Wandering flushes a glory that fades with arrival.

I came late to the love of birds. For years I saw them only as a tremor at the edge of vision. They know suffering and joy in simple states not possible for us. Their lives quicken and warm to a pulse our hearts can never reach. They race to oblivion. They are old before we have finished growing.

... For ten years, I spent all my winters searching for that restless brilliance, for the sudden passion and violence that peregrines flush from the sky. For ten years, I have been looking upward for that cloud-biting anchor shape, that crossbow flinging through the air. The eye becomes insatiable for hawks. It clicks towards them with ecstatic fury, just as the hawk's eye swings and dilates to the luring food-shapes of gulls and pigeons.

To be recognized and accepted by a peregrine, you must wear the same clothes, travel by the same way, perform actions in the same order. Like all birds, it fears the unpredictable. Enter and leave the same fields at the same time each day, soothe the hawk from its wildness by a ritual of behavior as invariable as its own. Hood the glare of the eyes, hide the white tremor of the hands, shade the stark reflecting face, assume the illness of a tree. A peregrine fears nothing he can see clearly and far off. Approach him across open ground with a steady unfaltering movement. Let your shape grow in size but do not alter its outline. Never hide yourself unless concealment is complete. Be alone. Shun the furtive oddity of man, cringe from the hostile eyes of farms. Learn to fear. To share fear is the greatest bond of all. The hunter must become the thing he hunts. What is, is now, must have the quivering intensity of an arrow thudding into a tree. Yesterday is dim and monochrome. A week ago you were not born. Persist, endure, follow, watch.

—*J.A. Baker, from* The Peregrine

One July afternoon at our ranch in the Canadian Rockies I rode toward Helen Keller's cabin. Along the wagon trail that ran through a lovely wood we had stretched a wire, to guide Helen when she walked there alone, and as I turned down the trail I saw her coming.

I sat motionless while this woman who was doomed to live forever in a black and silent prison made her way briskly down the path, her face radiant. She stepped out of the woods into a sunlit open space directly in front of me and stopped by a clump of wolf willows. Gathering a handful, she breathed their strange fragrance: her sightless eyes looked up squarely into the sun, and her lips, so magically trained, pronounced the single word "Beautiful!" Then, still smiling, she walked past me.

I brushed the tears from my own inadequate eyes. For to me, none of this exquisite highland had seemed beautiful. I had felt only bitter discouragement over the rejection of a piece of writing. I had eyes to see all the wonders of woods, sky, and mountains, ears to hear the rushing stream and the song of the wind in the treetops. It took the sightless eyes and sealed ears of this extraordinary woman to show me beauty, and bravery.

—*Frazier Hunt, from* Redbook

Whose woods these are I think I know
His house is in the village though;
He will not see me stopping here
To watch his woods fill up with snow.

My little horse must think it queer
To stop without a farmhouse near
Between the woods and frozen lake
The darkest evening of the year.

He gives his harness bells a shake
To ask if there is some mistake.
The only other sound's the sweep
Of easy wind and downy flake.

The woods are lovely, dark and deep.
But I have promises to keep,
And miles to go before I sleep,
And miles to go before I sleep.

—Robert Frost, from
"Stopping by Woods on a Snowy Evening"

The lesson in running brooks is that motion is a great purifier and health producer. When the brook ceases to run, it soon stagnates. It keeps in touch with the great vital currents when it is in motion, and unites with other brooks to help make the river. In motion it soon leaves all mud and sediment behind. Do not proper work and the exercise of willpower have the same effect upon our lives?

The other day in my walk, I came upon a sap bucket that had been left standing by the maple tree all the spring and summer. What a bucketful of corruption was that, a mixture of sap and rain-water that had rotted and smelled to heaven. Mice and birds and insects had been drowned in it, and added to its unsavory character. It was a bit of Nature cut off from the vitalizing and purifying chemistry of the whole. With what satisfaction I emptied it upon the ground while I held my nose and saw it filter into the turf, where I knew it was dying to go and where I knew every particle of the reeking, fetid fluid would soon be made sweet and wholesome again by the chemistry of the soil.

—*John Burroughs, from* The Gospel of Nature

The sounding cataract
Haunted me like a passion: the tall rock,
The mountain, and the deep and gloomy wood,
Their colors and their forms, were then to me
An appetite; a feeling and a love.

—*William Wordsworth*

The Season of Trilliums

A fugitive of the work ethic,
I step into woods sweetened
by the smell of fresh verdure
and the first lilies flowering.
Trilliums hide in plain view
as the sun burns through
naked branches, scorching
the crackled leaf litter underfoot.

Skirting a pond
rippled by jumping fish,
I listen to the sounds of renewal —
to peepers, to warblers and
woodpeckers knocking.
Last night I stayed awake
as the blackened sky flashed daylike
amid peals of thunder, wondering
how long happiness lasts.
Dawn came much faster than expected.

Now it's clear, at midday,
in my German shepherd's unflinching stare
as she stands chest deep in the water,
that time and joy are as fleeting
as the winged life of a mayfly.
So I revel in this moment
between anticipation and memory,
silently singing the glories
of mud, sweat and bloodletting flies.

—*Walt McLaughlin*

Bright Angel Point at Sunset

Thus the light rains, thus pours,
The liquid and rushing crystal
beneath the knees of the gods.
—Pound, Canto IV

The canyon bleeds, then deepens
and darkens. The intricate declension
of its ledges, bluffs and grottos
blends in this late light.
Wind swirls from the depths
carrying pine scent on its back.
A sliver of white moon
in the east. A nighthawk roars above.
Thin light spills into the gorge
and the river sings an ancient song.
At the edge of shadow, night:
dark stone, pine scent, water, cascading light.
—David Lee, from So Quietly the Earth

Dinosaurs died some 65 million years ago in the great worldwide Cretaceous extinction that also snuffed out about half of the species of shallow water marine invertebrates. They had ruled terrestrial environments for 100 million years and would probably rule today if they had survived the debacle. Mammals arose at about the same time and spent their first 100 million years as small creatures inhabiting the nooks and crannies of a dinosaur's world. If the death of dinosaurs had not provided their great opportunity, mammals would still be small and insignificant creatures. We would not be here, and no consciously intelligent life would grace our earth. Evidence gathered since 1980 indicates that the impact of an extraterrestrial body triggered this extinction. What could be more unpredictable and unexpected than comets or asteroids striking the earth literally out of the blue? Yet without such impact, our earth would lack consciously intelligent life. Many great extinctions (several larger than the Cretaceous event) have set basic patterns in the history of life, imparting an essential randomness to our evolutionary pageant.

—*Stephen Jay Gould, from* The Flamingo's Smile: Reflections in Natural History

By June 21, I had been living in my cottage for one year. It rained that night, a warm sustaining rain that dripped off the leaves in the hickory grove and filtered down through the tangle of wildflowers into the soil of the meadow. Just before going to bed, I went out and stood in the open air, allowing the cleansing coolness of the sky to fall over my shoulders. I was alone, and below the meadow, in my old house, a light was burning, a brighter reflection of the warmer light of the oil lamps in my cottage. I thought of a flicker I had heard the night before. For some unknown reason, in the middle of the night, it had let out a long whinny from the woods beside the cottage. The sound woke me instantly, and I felt a strange sense of communion with the bird — a fellow traveler in the experiment of life, a spark in a generally lifeless and desolate universe. I felt a similar communion seeing the light below the meadow. I felt that I and my family, my friends and allies and acquaintances, were all shrinking down into the small, wild spaces of the world. I was determined to stay on.

The rain slowed, spilled into a mere drip in the surrounding woods; a cricket started up, and deep in the mat of grasses on the south side of the meadow I saw the bright flash of a firefly.

—*John Hanson Mitchell, from* Living at the End of Time

You ask
why I perch
on a jade green mountain
I laugh
but say nothing
my heart
free
like a peach blossom
in the flowing stream
going by
in the depths
in another world
not among me.

—*Li Po, from* One Hundred Poems from the Chinese *translated by Kenneth Rexroth*

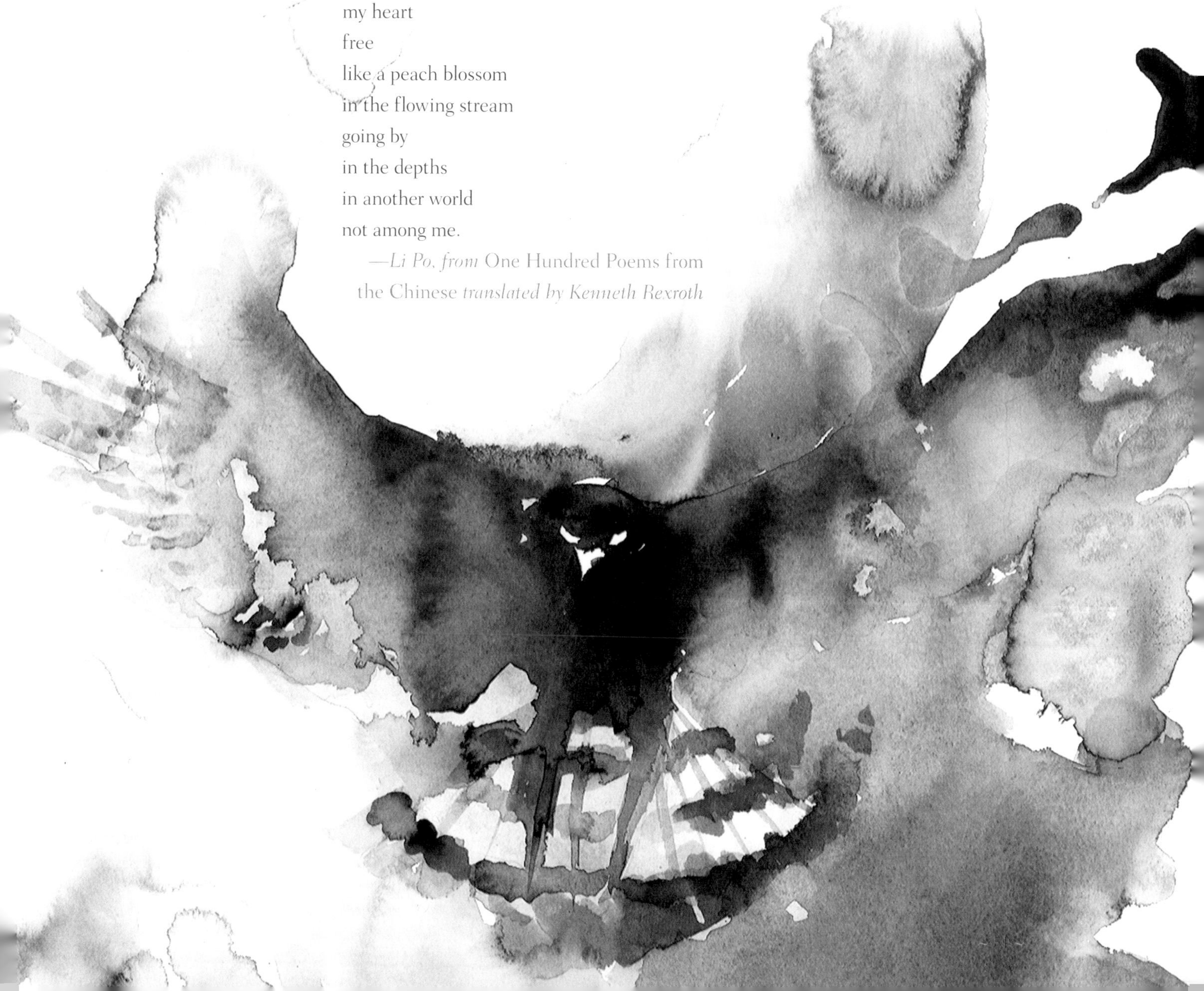

The vast majority of the roughly 33,500 species of spiders worldwide do indeed fly, although ballooning is a more accurate term (tarantulas and some other primitive species are among the few that do not balloon). There are different methods the spiders use to launch themselves into the air, but the most common way goes like this: the spider seeks a nearby high point by climbing onto a plant, fence post or other object, turns face first into the wind and firmly grabs the surface with all eight legs. The spider then raises its abdomen skyward and begins to pay out silk from its spinnerets and onto the wind. Even the mildest breeze can catch the line while giving assistance in pulling more silk out and onto the air currents. When enough silk for buoyancy is emitted, the spider releases its hold and is cast into the air....

Most spider flights are fairly short, but spiders have been captured by special aircraft as high as 14,000 feet and on ships 200 miles from the nearest shore. Spiders have been found at 22,000 feet on Mount Everest, apparently surviving on the dead bodies of tiny insects carried there by the winds, making them the largest — and perhaps only — animals that can live at the highest elevations on the earth's surface.

—*Dr. Robert Gale Breen III,* Backyard BUGwatching,
the magazine of the Sonoran Arthropod Studies Institute, 1992

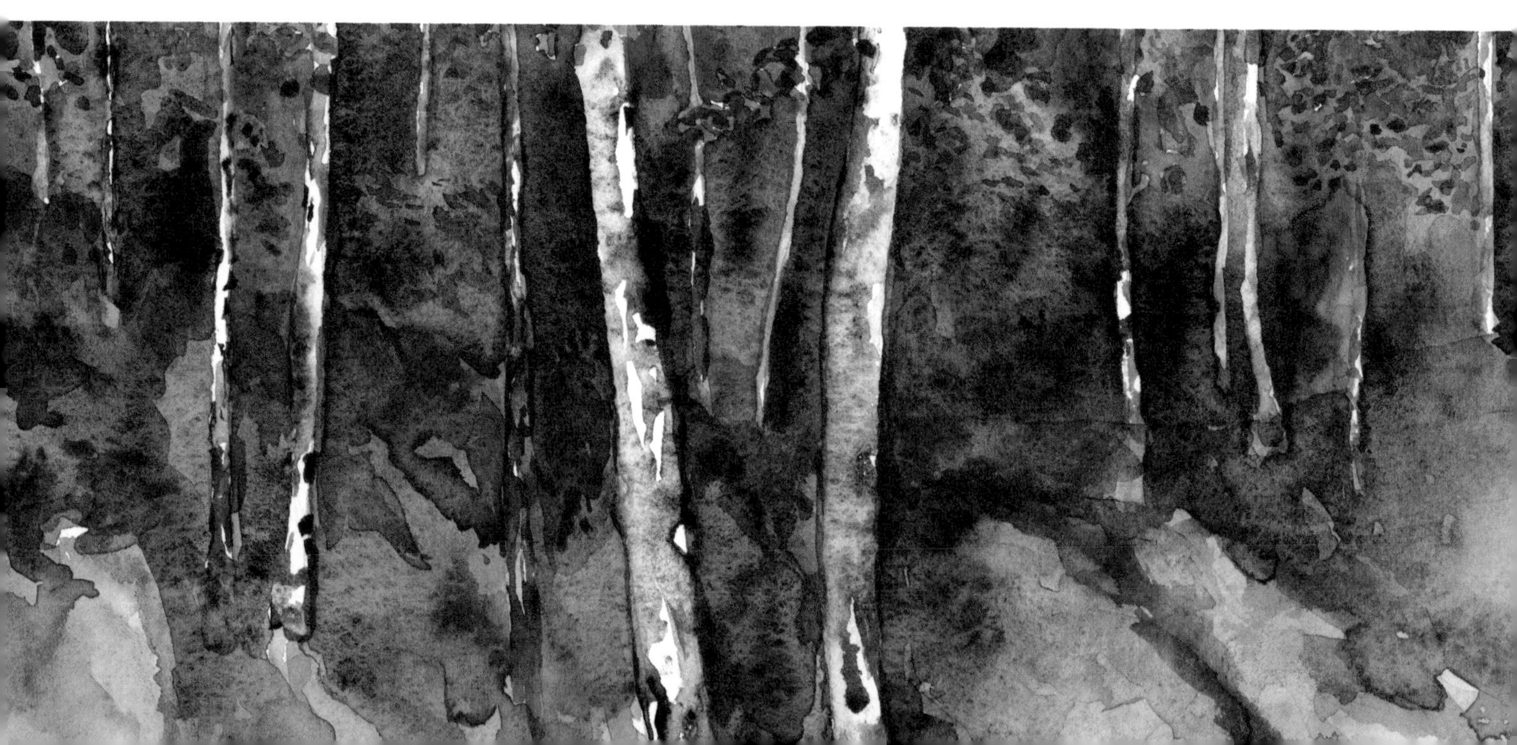

When all thoughts
Are exhausted
I slip into the woods
And gather
A pile of shepherd's purse.
Like the little stream
Making its way
Through the mossy crevices
I, too, quietly
Turn clear and transparent.

—*Ryokan*

We need the tonic of wildness, to wade sometimes in marshes where the bittern and the meadow-hen lurk, and hear the booming of the snipe; to smell the whispering sedge where only some wilder and more solitary fowl builds her nest, and the mink crawls with its belly close to the ground. At the same time that we are earnest to explore and learn all things, we require that all things be mysterious and unexplorable, that land and sea be infinitely wild, unsurveyed and unfathomed by us because it is unfathomable. We can never have enough of nature. We must be refreshed by the sight of inexhaustible vigor, vast and titanic features, the sea-coast with its wrecks, the wilderness with its living and its decaying trees, the thunder cloud, and the rain which lasts three weeks and produces freshets. We need to witness our own limits transgressed, and some life pasturing freely where we never wander.

—*Henry David Thoreau, from* Walden

I'm enjoying life more fully now than I ever have ... I don't for a moment regret being seventy-two years old. It's part of life, just like getting born was; just like being a jack-ass and an adolescent was. (Laughs.) And I'm continually, repeatedly discovering or having experiences sitting out in the yard and listening to the spring coming into the land, watching my purple martins, knowing that they've been all the way down to the southern tip of South America, have come back, found the same house — same hole in the house (laughs) — that they were raised in. This great capacity of life to renew itself. I think perhaps I'm more sensitive to that than ever before. I'm more sensitive to the fact that when Robert Browning had Rabbi Ben Ezra say, "Grow old along with me, the best is yet to be, the last of life, for which the first was made" — that this is absolutely profound.

—John Henry Faulk, radio broadcaster, "Johnny's Front Porch,"
from The Search for Meaning *by Philip L. Berman*

What shall we do with a man who is afraid of the woods, their solitude and darkness? What salvation is there for him? God is silent and mysterious.

—Henry David Thoreau, from a journal entry, 1850

How significant that the rich, black mud of our dead stream produces the water-lily, — out of that fertile slime springs this spotless purity! It is remarkable that those flowers which are most emblematical of purity should grow in the mud.

—Henry David Thoreau, from a journal entry

Silence is the universal refuge, the sequel to all dull discourses and all foolish acts, a balm to our every chagrin, as welcome after satiety as after disappointment

—Henry David Thoreau, from a journal entry

We are all joined in the holiness of the mind that God created. I therefore never consider myself alone in the silence. I'm 90 and live alone in a four room house surrounded by open space. I have no TV and have always luxuriated in silence. The exterior silence is here. The interior silence is a work in progress.

—Hazel, with thanks to Friends of Silence

Silence is the communing of a conscious soul with itself. If the soul attend for a moment to its own infinity, then and there is silence. She is audible to all men, at all times, in all places, and if we will we may always hearken to her admonitions.

—Henry David Thoreau, from a journal entry, 1838

The small truth has words that are clear; the great truth has great silence.

—Tagore

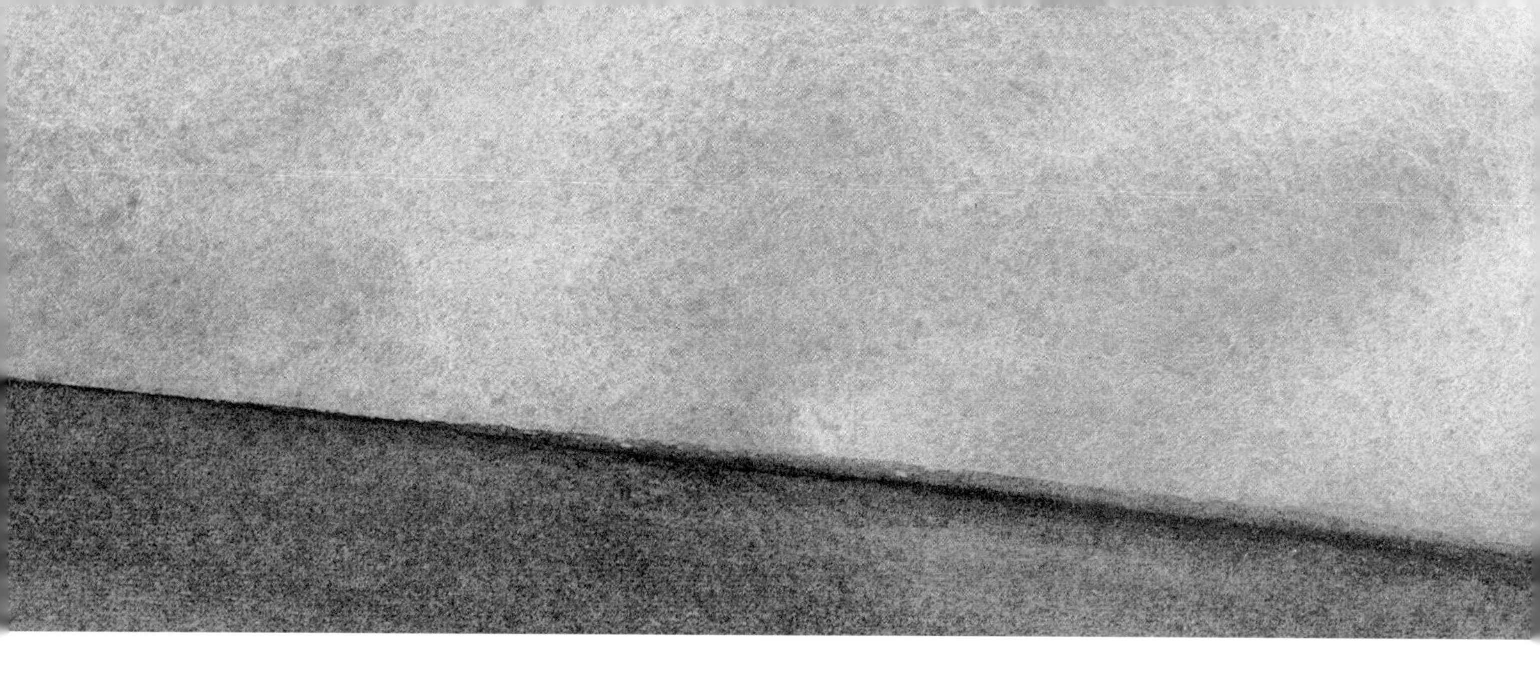

In silence we must wrap much of our life, because it is too fine for speech, because also we cannot explain it to others, and because some of it we cannot yet understand.

—Ralph Waldo Emerson

At times on quiet waters one does not speak aloud but only in whispers, for then all noise is sacrilege.

—Sigurd F. Olson

One time, we were paddling in the Baja, through some perfectly smooth, jade green water. On the bottom were sea cucumbers, rock scallops, some beautiful sea weeds. Everybody started to go very slow. We had been paddling hard, but everybody slowed down and just started looking. I got inspired to write some poetry, so I pulled out my notebook. I started to write. After almost fifteen minutes of quiet, we picked up our paddles and started paddling again. There was so much power in that silence, that the next day we did the whole day in silence. Nobody talked at all. In silence there is a lot more access, I think, to the spirit, to spirituality, to the soul.

There is a little voice in all of us that is just a whisper. A tiny whisper. When you go into nature, into the wilds, especially alone, the whisper can come out and talk more. When you are in the city, you always have a list of things to do, to watch on TV, to think about. You can't listen to the whisper. But when you are outside, you have much less to distract you. Inside each of us is the spirit that whispers. This little voice is our true self. If we can listen, it will start to get louder. Eventually, that whisper will be our normal voice. That's when I really live. That is when dreams become reality. When I live from that deep intuitive place.

I want my work to come out of that place. If my work comes out of that place, that is the big shift. I used to think that if I did a certain kind of work, I could become a certain kind of person. Now I realize that I have to continually do whatever it is that will allow my work to come out of that deep place, where the poetry came out of when I was floating over the still water. If my work comes from there, then whatever I do is somehow going to be supported.

—Jennifer Hahn, Elakah Kayak Tours, from Issue 1 of Heron Dance

When I detect a beauty in any of the recesses of nature, I am reminded, by the serene and retired spirit in which it requires to be contemplated, of the inexpressible privacy of a life,—how silent and unambitious it is. The beauty there is in mosses will have to be considered from the holiest, quietest nook.

The gods delight in stillness; they say, 'St—'st. My truest, serenest moments are too still for emotion; they have woolen feet. In all our lives we live under the hill, and if we are not gone we live there still.

—Henry David Thoreau, from a journal entry, January 1841

Place

Inside an open rose
A tree frog
no bigger
than my thumbnail.
I try to imagine
rest like that,
tucked
in such a bed of petals.
I try to imagine
prayer like that,
listening
so intently
in the early light
and
saying so little.
The summer
Teeters
Now
into old age,
as do I,
those blackberries
that still cling
to their thorny arms
withering,
readying themselves
to trust the earth again,
where,
for a moment
at least,
there is a place for everything.

—Bernardo Taiz

It is a paradox that we encounter so much internal noise when we first try to sit in silence.

—*Gunilla Norris, with thanks to* Friends of Silence

Silence is a privileged entry into the realm of God and into eternal life. There is a huge silence inside each of us that beckons us into itself, and the recovery of our own silence can begin to teach us the language of heaven. For silence is a language that is infinitely deeper, more far-reaching, more understanding, more compassionate, and more eternal than any other language.... There is nothing in the world that resembles God as much as silence.

—*Meister Eckhardt, with thanks to* Friends of Silence

When I Heard the Learn'd Astronomer

When I heard the learn'd astronomer,
when the proofs, the figures, were ranged in columns before me,
When I was shown the charts and diagrams, to add, divide, and measure them,
When I sitting heard the astronomer where he lectured with much applause in the
lecture-room,
How soon unaccountable I became tired and sick,
Till rising and gliding out I wander'd off by myself,
In the mystical moist night-air, and from time to time,
Look'd up in perfect silence at the stars.

—*Walt Whitman*

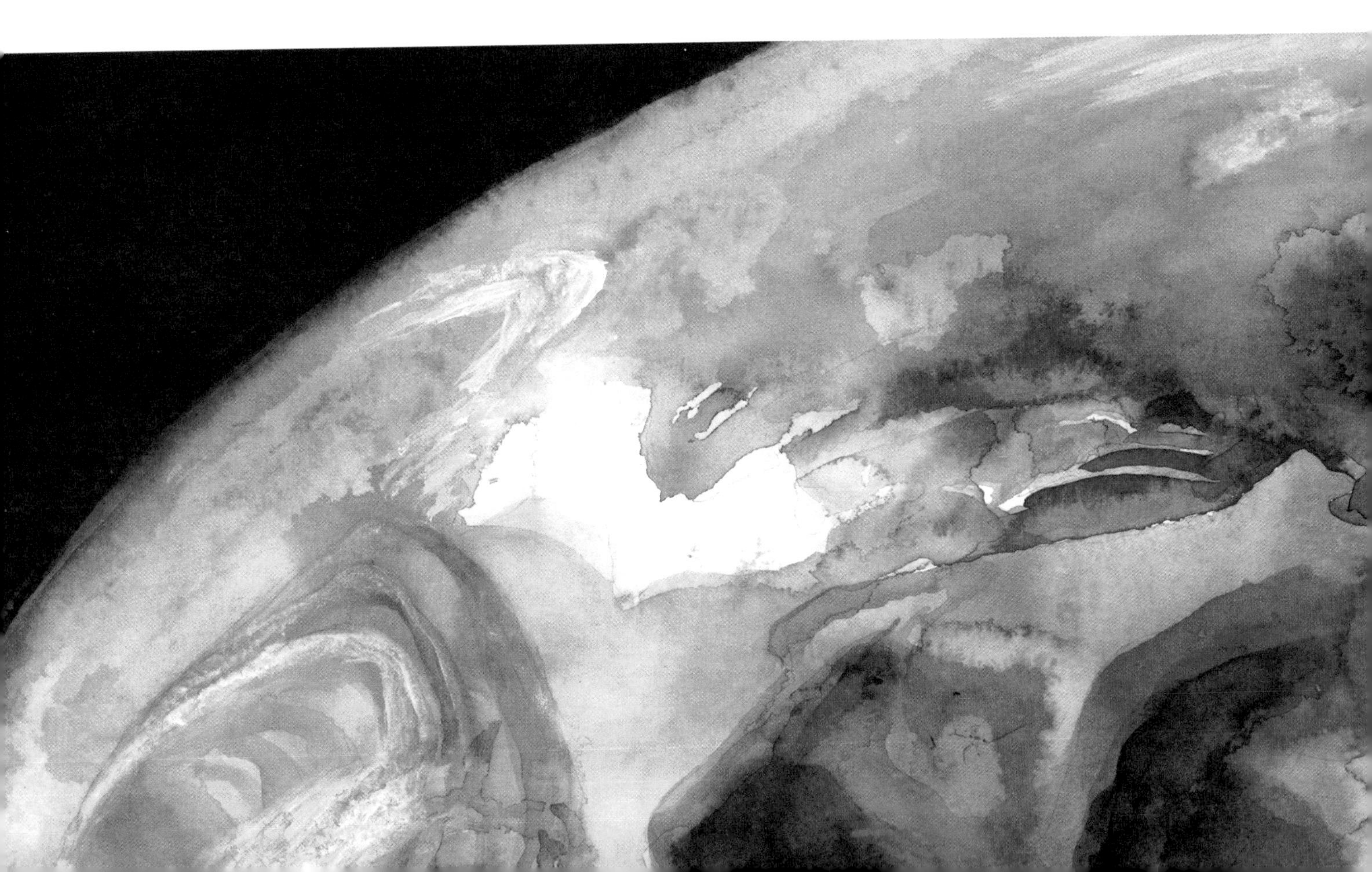

Forests and fields, sun and wind and sky, earth and water, all speak the same language: peace, solitude, silence.

—*Thomas Merton*

There are voices which we hear in solitude, but they grow faint and inaudible as we enter into the world.

—*Ralph Waldo Emerson*

Days and months are the travelers of eternity. So are the years that pass by.... I myself have been tempted for a long time by the cloud-moving wind — filled with a strong desire to wander ... I walked through mists and clouds, breathing the thin air of high altitudes and stepping on slippery ice and snow, till at last through a gateway of clouds, as it seemed, to the very paths of the sun and moon. I reached the summit, completely out of breath and nearly frozen to death. Presently the sun went down, and the moon rose glistening in the sky.

—*Matsuo Basho, from* The Narrow Road to the Deep North

Wilderness should be sacred and quiet, just as the Indians felt in designating certain places as spirit lands where no one talked. I have written about the Kawashaway River country of "no place between," where the Indians always traveled quietly and spoke only in whispers ... two of the greatest values of wilderness travel, solitude and silence.

—*Sigurd F. Olson, from* Reflections from the North Country

As the sun set, the wind died. The evening became quiet. The last light had the same effect as a snowfall; a stillness settling over everything. I could hear the hum of mosquitoes on the other side of the river, the occasional plop of a trout surfacing. No deafening waterfalls here.

I climbed the esker in the tracks of the caribou to watch the sun set. On top, I was two hundred feet above the river. Nothing nearby was half as high. What an inversion from the Palmer Valley! The sun set, but the light lingered on different spots as if reluctant to leave them. The river below me became a sheet of iridescent, twisted sliver laid on the darker landscape. Off to my right, I could see black spots on the silver. They were the rocks of the river's first rapids. The evening's stillness allowed the rapids' deep rumble to drift up to me on the esker. The calmness of the evening settled in my stomach next to the trout.

The silence of the rocks is staggering. In a forest, even in a field, I am distracted from noticing them by the movement of the grass, or the trees. Activity and the earth often hid them and their beautiful, endless silence.

—Robert Perkins, from Against Straight Lines

The longest silence is the most pertinent question most pertinently put. Emphatically silent. The most important question, whose answers concern us more than any, are never put in any other way.

—Henry David Thoreau, from a journal entry, 1851

Dawn Psalm, Pine Valley

1
While I was not watching
sunrise came with a ruby throat
and gold-flecked wings.

2
Blue
and a small wisp of cloud
above the dark pine.
A jaysquall
leaves a bruise
on one corner
of sky.

3
Boiling coffee.
A blue enameled pot
nestled in warm coals
beside the cold
sliding water.
Sky so close
you fear
bumping your head.

4
A brown breaks surface
rising to wingshadow
drifting on the blue selvage
of pond.

5
Golden lace.
Sunrise pours slantwise
into clear water
through the blue spruce,
the deep tangle of pine
and purled woodsmoke.

6
I turned
and the earth
hushed.
While I leaned into silence
a morning too vast to fathom
filled with light.

7
Praise.

—David Lee, from
So Quietly the Earth

The Lake Isle of Innisfree

I will arise and go now, and go to Innisfree
And a small cabin build there, of clay and wattles made
Nine bean rows will I have there, a hive for the honeybee
And live alone in the bee-loud glade.

And I shall have some peace there, for peace
Comes dropping slow.
Dropping from the veils of the morning to where the cricket sings;
There midnight's all a glimmer, and noon a purple glow,
And evening full of the Linnet's wings.

I will arise and go now, for always night and day
I hear lake water lapping with low sounds by shore;
While I stand on the roadway, or on the pavements grey,
I hear it in the deep heart's core.

—*William Butler Yeats, 1890*

I want you to understand something about long-distance walking. It's not a physical kind of thing. Oh, there's the rhythm, the sweat, the power of your legs chugging away. That's a thing most of us can comprehend, and this is indeed very physical, but most folks think of these long hikes a feat of some super being. If you could see me, you'd know the truth. I'm tall and skinny. I'm not an athlete. I'm not in any way, shape or form a super hiker. The physical part of the trek is real and demanding, but more important, I think, is the will. When it comes right down to it, walking is often a state of mind. Many days are a trial for me physically, yet because of my intense desire to be there, the miles drift by undetected. Often, it seems, sheer willpower pulls me over a mountain.

As in anything, one must be intent upon the path. When you're unhappy about The Way, that's when The Way becomes difficult. A long trek is a journey through the land, but it is also a searching odyssey into your self. It is not for everyone. To be alone and facing the wild places on their own terms is often easy compared to looking into our true selves for weeks and months at a time, which being alone on the trail forces us to do. But it is often the connection with the wilderness that pulls us through the confrontations within ourselves. I wouldn't want to be alone for so long in any other kind of environment.

But these are just outlines. What of the actual steps? When does walking, the motion, become thought? This is what I want you to understand. There is a point where the action goes unrecognized and is replaced by an emptiness that is based totally in the present. It is a knowledge, gained in each step, flooded with each turn, and brought into being for each second of the day. It is a holy state. This is when I wish I could tap myself, pouring my essence into a bottle. When someone asks me, "Why do you walk?" I'd just pull out the brew and say, "Drink this." Some days I can feel the miles building up under me, piling up and spilling over. But then there are days, like this one, when I float in a blessed euphoria. Steps cease to be merely feet gained. Rather, each action is a journey of its own. Each step holds its own wonders. And joy floods me at every turn.

—*Walkin' Jim Stoltz, from* Walking with the Wild Wind

Out of the clouds I hear a faint bark, as of a faraway dog. It is strange how the world cocks its ear to that sound, wondering. Soon it is louder: the honk of geese, invisible, but coming on.

The flock emerges from the low clouds, a tattered banner of birds, dipping and rising, blown up and blown down, blown together and blown apart, but advancing, the wind wrestling lovingly with each winnowing wing. When the flock is a blur in the far sky I hear the last honk, sounding taps for summer.

It is warm behind the driftwood now, for the wind has gone with the geese. So would I — if I were the wind.

—*Aldo Leopold, from* Sand County Almanac

At this time in the evening we would see the stars begin to appear as the sun disappears over the horizon. The light of day gives way to the darkness of night. A stillness, a healing quiet comes over the landscape.

It's a moment when some other world makes itself known, a numinous presence beyond human understanding. We experience the wonder of the things as the vast realms of space overwhelm the limitations of our human minds. As the sky turns golden and the clouds reflect the blazing colors of evening, we participate for a moment in the forgiveness, the peace, the intimacy of all things with each other.

—Thomas Berry, from the lecture "The Great Community of Earth," United Nations Millenium World Peace Summit, August 30, 2000

The three months of solitude were of the greatest significance for me. I came away from them very much strengthened — ready to share my insights with people who were interested in hearing them. Still today, I have the sounds of the jungle in my ears: the cries of the monkeys and birds and the wind rushing through the banana leaves. But there were also times of utter silence, at dawn and twilight. I took walks in the jungle in order to look at nature as a part of myself.

—*Ayya Khema, from* I Give You My Life

What did you find in the fields today,
you who have wandered so far away?
I found a wind-flower, small and frail,
and a crocus cup like a holy grail;
I found a hill that was clad in gorse,
a new-built nest, and a streamlet's source;
I saw a star and a moonlit tree;
I listened ... I think God spoke to me.

—*Hilda Rostron, with thanks to* Friends of Silence

To live a contemplative life is to be open enough to see, free enough to hear, real enough to respond. It is a life, and so it has its own rhythms of darkness, of dying-rising. Simply enough, it is a life of grateful receptivity, of wordless awe, of silent simplicity.

—*S. Marie Baha, with thanks to* Friends of Silence

Only in silence the word,

only in dark the light,

only in dying life.

—*Ursula K. LeGuin, from* A Wizard of Earthsea

Anyone who has probed the inner life, who has sat in silence long enough to experience the stillness of the mind behind its apparent noise, is faced with a mystery. Apart from all the outer attractions of life in the world, there exists at the center of human consciousness something quite satisfying and beautiful in itself, a beauty without features. The mystery is not so much that these two dimensions exist — an outer world and the mystery of the inner world — but that we are suspended between them, as a space in which both worlds meet … as if the human being is the meeting point, the threshold between two worlds.

—*Kabir Helmisnski, from* The Knowing Heart

What fascinates me the most are the cycles. Millions and billions of herring just came from places we don't even know migrate into all of these little bays; in the same bays they've been going into for thousands of years. Little genetically unique species of herring going into the same little bay where they've laid all of their eggs, sprayed all their sperm in the water and billions upon billions of little eggs are now forming little eyeballs right now and are about to sprout into fry. The sea lions and whales follow them in and … an incredible cycle. These are the kind of life processes that fascinate me and, I think, that's at the core of what we're trying to protect and what keeps us going. That humbling notion that we're just scratching the surface of our understanding of these systems and yet humans are manipulating and changing them to such a profound degree.

—Ian McAllister, Raincoast Conservation Society,
from a Heron Dance *interview, www.herondance.org*

Once in a lifetime, perhaps, one escapes the actual confines of the flesh. Once in a lifetime, if one is lucky, one so merges with sunlight and air and running water that whole eons, the eons that mountains and deserts know, might pass in a single afternoon without discomfort. The mind has sunk away into its beginnings among old roots and the obscure tricklings and movings that stir inanimate things. Like the charmed fairy circle into which a man once stepped, and upon emergence learned that the whole century had passed in a single night, one can never quite define this secret; but it has something to do, I am sure, with common water. Its substance reaches everywhere; it touches the past and prepares the future; it moves under the poles and wanders thinly in the heights of the air. It can assume forms of exquisite perfection in a snowflake or strip the living to a single shining bone cast up by the sea.

Loren Eisley, from The Immense Journey

If waters are placid, the moon will be mirrored perfectly. If we still ourselves, we can mirror the divine perfectly. But if we engage solely in the frenetic activities of our daily involvements, if we seek to impose our own schemes on the natural order, and if we allow ourselves to become absorbed in self-centered views, the surface of our waters becomes turbulent. Then we cannot be receptive to Tao.

There is no effort that we can make to still ourselves. True stillness comes naturally from moments of solitude where we allow our minds to settle. Just as water seeks its own level, the mind will gravitate toward the holy. Muddy water will become clear if allowed to stand undisturbed, and so too will the mind become clear if it is allowed to be still.

Deng Ming-Dao, from 365 Tao: Daily Meditations

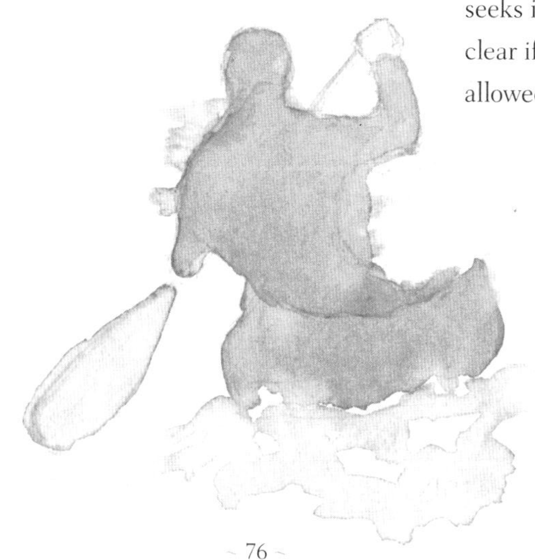

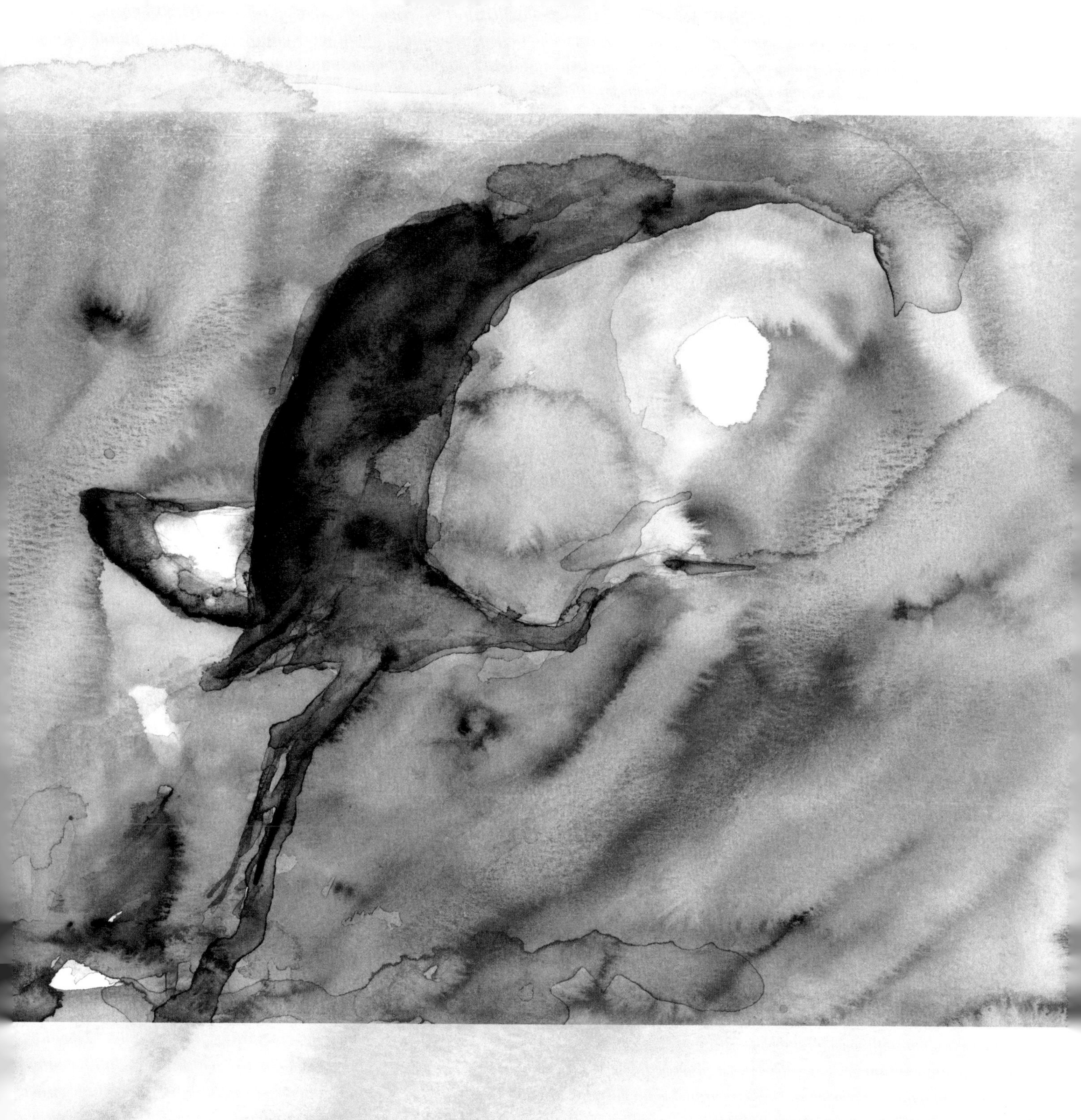

The surrounding forest looks familiar, but I'm not quite in sync with it yet. A part of me is still in a Williamstown café, enjoying brunch with my wife, Judy, while listening to classical music. It will take a while to match the natural rhythm of things. Several thresholds have to be crossed before that can happen. The first will come in a matter of hours, another after a night in the woods, yet another in a few days. I relish this gradual return to a simpler existence, yet something deep inside me recoils in the mute terror from it. More has been abandoned at the trailhead than the mere amenities of modern living. With a little coaxing from the forest, a wilder self will slowly emerge but only at the expense of something more refined, genteel. It's a trade-off to be sure.

... Steady movement is much like a mantra chanted over and over. Steady movement reduces inner confusion, and chaos, the same way that white noise drowns out the invasive, distracting sounds of daily life. I immerse myself in the Eternal Present as I walk, reeling in the absolute immediacy of a trail twisting and turning through the trees. The forest opens to me. Around every bend, a new world awaits—a glimpse of some wild creature darting across the land, a beaver pond appearing unexpectedly, a rare flower blooming. There is no end to it. I savor these small, elemental surprises as the abstract concerns of a more complicated way of life gradually fade away. I become a part of the forest—a woods wanderer, a seeker of wild things, a chaser of butterflies. And the gap between self and other narrows.

—*Walt McLaughlin, from* Forest Under My Fingernails
(*available through* Heron Dance)

Algonquin Loons by Roderick MacIver

Blue sky. I am eight years old, lying on my back on the ground in the woods. I lie there and stare for two or three hours at the mashed potato clouds and their changing forms as they drift by overhead.

One afternoon, about twenty-five years later, I was driving north to canoe for a week in Canada's Algonquin Park. I stopped by my parents' house in Ottawa, and as I was leaving, I overheard my dad say to one of his old army buddies, "He lives for these trips." That surprised me. I am not even sure I had yet come to completely realize how much those one-week solo trips, which I found time to take only every year or two, had come to be the center of my life. I was surprised he had seen that. I got in my car, waved, and drove off. I have never mentioned it to him, but I think often of that offhand comment of his.

On the surface, I was in my early thirties, had a young family, and was working in the investment business. On the surface, my dad was an ex-soldier, a bureaucrat in the secret police. We've had lots of ups and down, my dad and I, and yet maybe we understand more about each other than we let on.

Thirty-six or so hours later, I was sitting beside a lake that was shrouded in a gray, pre-dawn fog. I had gotten up in the dark, packed my canoe, and was sitting out at the shore sipping tea. Two loons drifted in and out of view. One stretched and called that haunting call of the wild.

Four years later, I had to spend one week every couple of months in a hospital getting experimental chemotherapy. I would lie in the hospital bed, close my eyes, and see and hear that lake and those loons. The actual experience lasted perhaps twenty seconds, but it sustained me through hours and months.

Perhaps each human life is fed by the underground spring of a few experiences. When we are there, we touch something beyond words. Four such experiences come to my mind. Perhaps there have been a few more. They make me who I am to me, who I am under the personas I assume to negotiate my way in the world.

It amazes me how often in my life I have embarked upon work, upon commitments that have absorbed years of effort, that have nothing whatever to do with those moments of deep peace and joy. In fact, I can say that I have spent most of my life living as if I was trying to prove that those moments don't matter. As I approach the age of fifty, I see my life as a gradual reorientation, a gradual shifting of course, towards serving, with my art and words, a very few experiences.

List of Illustrations

All the watercolors in this book are by Roderick MacIver, founder of *Heron Dance*. Select watercolors from this list are available as full-color, limited-edition prints on the *Heron Dance* website (www.herondance.org) by typing the title in the search bar. If you have trouble finding the image you would like or do not have access to the internet, please call *Heron Dance* toll free at 888-304-3766 or send an email to heron@herondance.org. Thank you.

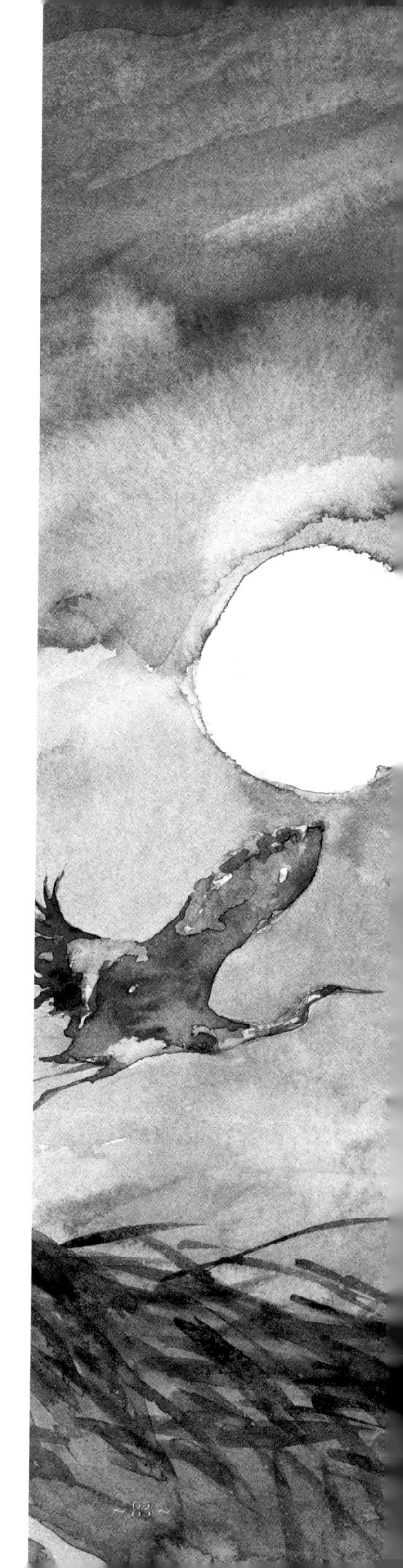

Heron Dance Press & Art Studio is a nonprofit 501(c)(3) organization founded in 1995. It is a work of love, an effort to produce something that is thought-provoking and beautiful. Through our website, quarterly journal, workshops, free weekly e-newsletter and watercolors, *Heron Dance* celebrates the seeker's journey and the beauty and mystery of the natural world.

We invite you to visit us at www.herondance.org to view the many beautiful watercolors by Roderick MacIver and to browse the hundreds of pages of book excerpts, poetry, essays, and interviews of authors and artists and to learn more about the following:

Nonprofit Donations
Heron Dance donates thousands of notecards and the use of hundreds of images to small nonprofits every year. Please contact us for more information.

Heron Dance* Journal and *A Pause for Beauty
Available by donation, *Heron Dance* is a 16-page, full-color journal published four times a year, featuring nature watercolors by Roderick MacIver, offering thought-provoking essays, interviews, poetry, and quotations centered around the human search for meaning and the human connection to the natural world. *Heron Dance* is a celebration of the gift of life!

A Pause for Beauty is a free daily email of watercolor art, and a poem, quotation, or reflection. To sign up or view our archives, visit our website.

Watercolors by Roderick MacIver
Hundreds of nature watercolors are available as signed limited-edition prints and originals at www.herondance.org.

Studio Store
We offer notecards, daybooks, calendars, address books, and blank journals that feature Roderick MacIver watercolors, along with inspirational titles from *Heron Dance Press*, including *The Heron Dance Book of Love and Gratitude*.

Heron Dance Community
The *Heron Dance* community consists of over 30,000 subscribers to our print journal *Heron Dance* and over 20,000 readers of our free daily email *A Pause for Beauty*. A community site is planned for completion in late 2006 that will allow our subscriber members to get in touch with each other for everything from discussing books and films to dating. To learn more, visit our website and enter "*Heron Dance* Dreams" in the search bar.

Open Heart ~ Wild Soul Workshops
These weekend retreats, led by *Heron Dance* editor and writer Ann O'Shaughnessy, are opportunities to gather with kindred souls to explore and nourish our truth in a tender and open-hearted way. Ann also publishes a twice-monthly e-newsletter called *Letters from an Open Heart* in which she offers inspiration and tools for the journey.

ALSO FROM HERON DANCE PRESS

All titles feature Roderick MacIver watercolors

Heron Dance *Book of Love and Gratitude*

Heron Dance celebrates the open heart and the beauty and mystery that surround us with this book of poetry, book, and interview excerpts. Designed by Ann O'Shaughnessy, it contains 48 watercolors by Roderick MacIver and selections from the written works of Helen Keller, Dostoevsky, and Henry Miller, among many others. 80 pages.

#1602 Heron Dance *Book of Love and Gratitude* — $12.00

Art as a Way of Life

This beautiful and inspiring book helps us to discover and nurture the creative spirit that is within us all. Filled with the beautiful nature watercolors of Roderick MacIver, Ann O'Shaughnessy's own short reflections, and selections from many other creative voices, *Art as a Way of Life* is a truly empowering and encouraging book for people wanting to live, work, and love in the creative spirit. 96 pages.

#6093 *Art as a Way of Life* — $12.95

Pausing for Beauty — The Heron Dance *Poetry Diary*

We designed this diary just as we wanted it—with plenty of inspiring poetry and art AND plenty of blank and lined pages. Containing over 140 of Rodericks nature watercolors, there are 12 full-page dateless calendars at the start of every month so that the book can be used for any year. 176 pages.

#1606 *Pausing for Beauty — The* Heron Dance *Poetry Diary* — $15.95

Thoreau and the Art of Life

Precepts and Principles

Henry David Thoreau wrote beautifully about nature, love and friendship, art and creativity, spirituality, aging and death, simplicity, and wisdom. This book draws from Thoreau's journals to offer unusual insights into Thoreau's deep reflections on living a life of meaning, and reverence. His powerful body of work continues to inspire us. 112 pages, full-color watercolors throughout.

#6091 *Thoreau and The Art of Life* — $12.95

The Laws of Nature

Excerpts from the Writings of Ralph Waldo Emerson

Emerson was among the first to incorporate the power of wild nature into his worldview, thus giving birth to Transcendentalism. Walt McLaughlin has compiled this selection of Emerson's most thought-provoking nature writing, gleaned from personal journals as well as published works, providing a glimpse into the mind of a true lover of nature. New expanded edition, 96 pages, full-color watercolors throughout.

#6092 *The Laws of Nature* — $12.95

Earth, My Likeness

Nature Poetry of Walt Whitman

Walt Whitman's love for wild nature and for the sensual experiences of life is heard in all his poems. Howard Nelson provides an insightful introduction, shedding light on Whitman's life. This A carefully selected collection of poems alongside Roderick MacIver's watercolor art creates a grand tribute. Revised edition, 144 pages.

#6088 *Earth, My Likeness* — $11.95

Forest Under My Fingernails

Reflections and Encounters on The Long Trail

Walt McLaughlin writes beautifully about his encounters and reflections during a month-long backpacking trek. We regard Walt as one of the most eloquent nature authors writing today. 192 pages.

#6085 *Forest Under My Fingernails* — $15.95

This Ecstasy

This courageous and beautiful book of poems by John Squadra explores, with simplicity, the truths of love and a spiritual life. Some poems are very erotic. Some poems expose the truths of life we all share. 96 pages.

#6087 *This Ecstasy* — $10.95